SCIENCE
and the Akashic Field

"Ervin Laszlo presents readers with a tour de force, nothing less than a theory of everything. This book introduces such provocative concepts as the "A-field" and the "informed universe," making the case that a complete understanding of reality is woefully lacking without them. Readers of this book will never view the universe in quite the same way again."

STANLEY KRIPPNER, PH.D.,
PROFESSOR OF PSYCHOLOGY, SAYBROOK GRADUATE SCHOOL,
AND AUTHOR AND CO-EDITOR OF *VARIETIES OF ANOMALOUS EXPERIENCE*

"Over the last 30 years, Ervin Laszlo has consistently been at the forefront of scientific inquiry, exploring the frontiers of knowledge with insight, wisdom and integrity. With *Science and the Akashic Field* he takes another quantum leap forward in our understanding of the universe and ourselves. This enthralling vision of mind, science, and universe is essential reading for the 21st century."

ALFONSO MONTUORI, PH.D.,
CALIFORNIA INSTITUTE OF INTEGRAL STUDIES,
AND AUTHOR OF *CREATORS ON CREATING*

"It is rare indeed that a revolution in thought can open our eyes to a new universe that transforms our inner experience as well as our relationships with others and even with the cosmos. Martin Buber did it with *I and Thou*. Now, Ervin Laszlo, one of the most profound minds of our generation, has given us a great gift in this readable book that explores how we are connected to each other in fields of resonance that penetrate to the deepest levels of being."

ALLAN COMBS, PH.D.,
PROFESSOR OF PSYCHOLOGY, UNIVERSITY OF NORTH CAROLINA AT ASHEVILLE,
AND AUTHOR OF *THE RADIANCE OF BEING*

"Laszlo's book opens the way toward a great synthesis. Whoever reads Laszlo's book witnesses the greatest awakening of the human spirit. Not since Plato and Democritus has there been such a transformation in the history of thought!"

"In his admirable 40-year quest for an integral theory of everything, Laszlo has not restricted himself to physics but presented a coherent global hypothesis of connectivity between quantum, cosmos, life and consciousness. I cannot think of anyone else who is better prepared and more able, than the genuine post-modern Renaissance Man Laszlo, to offer a vision that is imaginative, but not imaginary, a vision where all things are connected with all other things and nothing disappears without a trace."

"Is everything that ever happened on this earth recorded in some huge, ultra-dimensional information bank? Are some of us occasionally able to tap into it with some facility, and perhaps all of us to some extent now and then during our lives? *Science and the Akashic Field* provides the pioneering scientific answer to these and many other fundamental questions our species faces at this critical time in human evolution."

"*Science and the Akashic Field* shows clearly that science is poised at the threshold of a new paradigm. The new vision offers humanity the perspective of more peace and security, not as an idealistic goal but as a reflection of reality."

"When in search of impacts or nuances useful in discovering and understanding the essential universe, Ervin Laszlo's brilliant new work, *Science and the Akashic Field*, surpasses previous explorations. The work opens a road to understanding the universe as an integrated entity, connecting science and consciousness, and recognizing the wholeness of the universe, life, and mind. This is a *"make-sense-of-the-complex"* opus, accessible to every reader."

A. HARRIS STONE, ED.D.,
FOUNDER OF THE GRADUATE INSTITUTE IN MILFORD, CONNECTICUT,
AND AUTHOR OF THE LAST FREE BIRD

"There is turmoil and excitement at the cutting edge of cosmology and related sciences. Ervin Laszlo, with his insightful and systems-oriented approach, charts a course through it all that is both truly radical and truly plausible. This is a solidly grounded vision of our cosmos, with perspectives that are wide and deep and have profound implications for all of us."

HENRIK B. TSCHUDI,
CHAIRMAN OF THE FLUX FOUNDATION, OSLO, NORWAY

"Ervin Laszlo is, arguably, the most profound thinker alive today."

LADY MONTAGU OF BEAULIEU,
FIRST AMBASSADOR OF THE CLUB OF BUDAPEST

SCIENCE

and the
Akashic Field

An Integral Theory of Everything

ERVIN LASZLO

Inner Traditions
Rochester, Vermont

Inner Traditions
One Park Street
Rochester, Vermont 05767
www.InnerTraditions.com

LIBRARY OF CONGRESS CATALOGING-IN-PUBLICATION DATA
Laszlo, Ervin, 1932–
Science and the Akashic field : an integral theory of everything / Ervin Laszlo.
p. cm.
Includes bibliographical references and index.
ISBN 1-59477-042-5 (pbk.)
1. Akashic records. 2. Parapsychology and science. I. Title.
BF1045.A44L39 2004
501—dc22
2004016393

Printed and bound in the United States at Capital City Press

10 9 8 7 6 5 4 3 2

Text design by Rachel Goldenberg
Text layout by Virginia Scott Bowman
This book was typeset in Sabon with Avenir as a display typeface

---

*for Christopher and Alexander, who
continue to comprehend, connect,
and co-create—with love*

Contents

———

Akasha *(ā·kā·sha) is a Sanskrit word meaning "ether": all-pervasive space. Originally signifying "radiation" or "brilliance," in Indian philosophy* akasha *was considered the first and most fundamental of the five elements—the others being* vata *(air),* agni *(fire),* ap *(water), and* prithivi *(earth). Akasha embraces the properties of all five elements: it is the womb from which everything we perceive with our senses has emerged and into which everything will ultimately re-descend.* The Akashic Record *(also called* The Akashic Chronicle*) is the enduring record of all that happens, and has ever happened, in space and time.*

Acknowledgments

This book is the fruit of over forty years of seeking a view of the world that is meaningful as well as embracing, rigorous, and yet simple. I cannot possibly thank by name all the people who have furnished information to me in my search or, even more important, provided encouragement and inspiration. Let me mention merely those who have been most directly instrumental in drafting and completing this, the most recent and perhaps the most definitive of the nearly half-dozen books I have devoted to this quest. I begin with my immediate family.

Living with a person who seems obsessed with working out and communicating an idea is not an easy matter; I am deeply grateful to my wife, Carita, for putting up with both my absences and my absent-mindedness during the long periods when I was drafting, redrafting, and elaborating this book. Without her support and constant loving presence, I could not have had the peace, and the peace of mind, to undertake this project.

Once again, I dedicate this book to our sons, Christopher and Alexander, for they continue to remain "plugged in" as I range over fields as varied as the problems of morality and sustainability in today's world and the explanation of the strange finding that all things in the universe are connected with all other things. Their encouragement, love, and support, unobtrusive yet ever present, has been a major factor in my venturing on terrains where most academics, not to mention angels, fear to tread. Let me note that Kathia, Alexander's "better half" and closest collaborator, and Lakshmi, Christopher's spouse and life companion, are part of this intimate group of comprehension and co-creation.

A special note of thanks is due to my good friend the brilliant Hungarian physicist László Gazdag. His pathbreaking theories and rich background knowledge in avant-garde physics were an invaluable asset. Another person whose friendship and support were vital to this undertaking is my Club of Budapest colleague, gifted healer, and life-long friend Maria Sági. Her practical work in local as well as nonlocal diagnosing and healing—from which both I and my whole family have benefited—helped me find the way to the informed universe and gave me assurance that it is the right one.

There have been numerous friends and colleagues in the academic community who have followed my work and provided useful, often vital, information. Many of them have commented on this work prior to its publication. Let me take this opportunity to express my thanks to them, and to note that those who are members of the General Evolution Research Group—among them Allan Combs and David Loye—have been especially helpful and supportive.

A small but intensely committed group of colleagues who became friends (although some I have not even met in person) has been instrumental in editing, producing, and publishing this book. It includes first of all Bill Gladstone, head of Waterside Productions, whom I have known for years and who during all this time has steadfastly maintained that this book is my real intellectual legacy—notwithstanding many other books he has helped me develop and publish. It has been nearly five years since we envisaged this project and without his friendly but decisive insistence that I should "lower the altitude" of its language so as to make it accessible to a wide public, it would not have been completed in its present, hopefully clear and reader-friendly, form. In this regard I acknowledge with thanks the expert help of former Random House editor Peter Guzzardi, who, over a period of well over a year, has reviewed my successive drafts and offered valuable suggestions.

The team at Inner Traditions International proved to be a major asset. Well beyond the usual tasks of editors and publishers, the members of this team, headed by publisher Ehud Sperling, demonstrated the kind of creativity and commitment that used to be legendary in the publishing world but is mostly lacking in today's high-pressure business environment. I am pleased to acknowledge the vision of acquisitions editor Jon Graham, who, having had a look at an advance draft of the manuscript at the 2003 Frankfurt Book Fair, immediately decided that he wanted to acquire it. It is likewise a pleasure to acknowledge the collaboration of managing editor Jeanie Levitan, who—in charge of coordinating the various steps in the production and publication of this volume—has been thoroughly committed and heartwarmingly helpful throughout.

If I left to the last Nancy Yeilding, my copy editor, it is because she has been the last person with whom I have collaborated in this venture. When she took the text in her hands, I was fairly convinced that it was in final shape, save some linguistic touch-ups. But Nancy has done wonders in restructuring it for improved logic in exposition and enhanced clarity in language. The text before the reader bears the mark of her creative ideas—deeply appreciated by its author.

AUGUST 2004

Introduction

There are many ways of comprehending the world: through personal insight, mystical intuition, art, and poetry, as well as the belief systems of the world's religions. Of the many ways available to us, there is one that is particularly deserving of attention, for it is based on repeatable experience, follows a rigorous method, and is subject to ongoing criticism and assessment. It is the way of science.

Science, as a popular newspaper column tells us, matters. It matters not only because it is a source of the new technologies that are shaping our lives and everything around us, but also because it suggests a trustworthy way of looking at the world—and at ourselves in the world.

But looking at the world through the prism of modern science has not been a simple matter. Until recently, science gave a fragmented picture of the world, conveyed through seemingly independent disciplinary compartments. Even scientists have found it difficult to tell what connects the physical universe to the living world, the living world to the world of society, and the world of society to the domains of mind and culture. This is now changing; ever more scientists are searching for a more integrated, more unitary world picture. This is true especially of physicists, who are intensely at work creating "grand unified theories" and "super-grand unified theories." These GUTs and super-GUTs relate together the fundamental fields and forces of nature in a logical and coherent theoretical scheme, suggesting that they had common origins.

A particularly ambitious endeavor has surfaced in quantum physics in recent years: the attempt to create a theory of everything—a "TOE." This project is based on string and superstring theories (so called because in these theories elementary particles are viewed as vibrating filaments or strings), and it uses sophisticated mathematics and multi-dimensional spaces to produce a single equation that could account for all the laws of the universe. However, the TOEs of string theorists are not the definitive answer to the quest for a unitary world picture, for they are not really theories of *every*-thing—they are at best theories of every *physical*-thing. A genuine TOE would include more than the mathematical formulas that give a unified expression to the phenomena studied in this branch of quantum physics; there is more to the universe than vibrating strings and related quantum events. Life, mind, and culture are part of the world's reality, and a genuine theory of everything would take them into account as well.

Ken Wilber, who wrote a book with the title *A Theory of Everything,* agrees: he speaks of the "integral vision" conveyed by a genuine TOE. However, he does not offer such a theory; he mainly discusses what it would be like, describing it in reference to the evolution of culture and consciousness—and to his own theories. An actual, science-based integral theory of everything is yet to be created.

As this book will show, a genuine TOE *can* be created. Although it is beyond the string and superstring theories, in the framework of which physicists attempt to formulate their own super-theory, it is well within the scope of science itself. The factor required to create a genuine TOE is not abstract and abstruse: it is *information*—information as a real and effective feature of the universe. Although most of us think of information as data or what a person knows, physicists and other empirical scientists are discovering that information extends far beyond the mind of an individual person or even all people put together. In fact, it is an inherent aspect of nature. The great maverick physicist David Bohm called it "in-formation," meaning a message that actually "forms" the recipient. In-formation is not a human artifact, not something that we produce by writing, calculating, speaking, and messaging. As ancient sages knew, and as scientists are now rediscovering, in-formation is produced by the real world and is conveyed by a fundamental field that is present throughout nature.

When we recognize that "in-formation" (which for the sake of simplicity we shall write as "information") is a real and effective factor in the universe, we rediscover a time-honored concept—the concept of a universe that is made up neither of just vibrating strings, nor of separate particles and atoms, but is instead constituted in the embrace of continuous fields and forces that carry information as well as energy.

This concept—which is thousands of years old and has cropped up again and again in the history of thought—merits being known. First, because the energy- and information-imbued "informed universe" is a meaningful universe, and in our time of accelerating change and mounting disorientation we are much in need of a meaningful view of ourselves and of the world. Second, because understanding the essential contours of the informed universe does not call for having a background in the sciences; they are readily comprehendible by everyone. And last but not least, because the informed universe is probably the most comprehensive concept of the world ever to come from science. It is, at last, a truly unified concept of cosmos, life, and mind.

Science and the Akashic Field is a nontechnical introduction to the informed universe, cornerstone of a scientific theory that will grow into a genuine theory of everything. It describes the origins and the essential elements of this theory and explores why and how it is surfacing in quantum physics and in cosmology, in the biological sciences, and in the new field of consciousness research. It highlights the theory's crucial feature: the revolutionary discovery that at the roots of reality there is an interconnecting, information-conserving and information-conveying cosmic field. For thousands of years, mystics and seers, sages and philosophers maintained that there is such a field; in the East they called it the Akashic Field. But the majority of Western scientists considered it a myth. Today, at the new horizons opened by the latest scientific discoveries, this field is being rediscovered. The effects of the Akashic Field are not limited to the physical world: the A-field (as we shall call it) informs all livings things—the entire web of life. It also informs our consciousness.

THE STRUCTURE OF THIS BOOK

Scientists have often ignored the question of meaning in regard to their theories, considering it a philosophical if not downright metaphysical appendage to their mathematical schemes. This has impoverished the discourse of science and has had a negative impact on society. In chapter 1 we raise the question of meaning in regard to science and discuss the relevance of an up-to-date scientific worldview for our time. The worldview most people consider scientific is an inadequate and in many respects obsolete view. This, however, can be remedied.

Chapter 2 lays the groundwork for an encompassing scientific theory that is both meaningful for laypeople and capable of responding to the problems encountered by scientists. We review the "paradigm-shift" that promises to lead science toward such a theory. The key element is the accumulation of puzzles: anomalies that the current paradigm cannot clarify. This drives the community of scientists to search for a more fertile way of approaching the anomalous phenomena.

Chapter 3 offers a concise catalog of the findings that puzzle scientists in diverse fields of inquiry. This demonstrates the basic fact that evidence for a fundamental insight about reality does not come from a single experiment, or even from a single field of inquiry. If the insight is truly basic, its traces should be encountered in practically all systematic investigations of scientific interest. Our catalog of puzzles shows that this is the case in regard to the unsuspected forms and levels of coherence that come to light in the physical world and in the living world, as well as in the world of mind and consciousness.

In chapter 4 we enter on the quest of identifying nature's information field and building it into the spectrum of scientific knowledge. We explore theories of the quantum vacuum—the zero-point energy field that fills all of cosmic space—and discuss how this intensely researched but as yet incompletely understood cosmic field could convey not only energy, but also information.

Chapter 5 returns to a discussion of the evidence for information in nature, examining in greater detail the puzzles of science and describing how innovative scientists attempt to cope with them. A more profound look at both the evidence and the hypotheses by which the evidence is interpreted is indicated, since the assertion that an information field underlies all things in the universe is a major claim and—while it is a perennial insight of traditional cosmologies—it is a radical innovation in the eyes of conservative mainstream scientists.

In chapter 6 we go a step further: we present the scientific basis of the A-field, the cosmic information field. This is the foundation of a theory that can clarify many of the hitherto puzzling yet fundamental features of quanta and galaxies, organisms and minds. The resulting "integral theory of everything" takes information as a fundamental factor in the world. It recognizes that ours is not just a matter- and energy-based universe, but rather an information-based "informed universe." On first sight the informed universe may appear to be a surprising universe, yet on a deeper look it proves to be familiar—perhaps surprisingly familiar. Intuitive people have always known that the real universe is more than a world of inert, nonconscious matter moving randomly in passive space.

In chapters 7 and 8 we explore the informed universe. We ask some of the questions thinking people have always asked about the nature of reality. Where did the universe come from? Where is it going? Is there life elsewhere in the wide reaches of the universe? If so, is it likely to evolve to higher stages or dimensions? We go on to ask questions about the nature of consciousness. Did it originate with *Homo sapiens* or is it part of the fundamental fabric of the cosmos? Will it evolve further in the course of time—and what kind of impact will it have on our world when it does?

We probe deeper still: Does human consciousness cease at the physical death of the body or does it continue to exist in some way, in this or in another sphere of reality? And could it be that the universe itself possesses some form of consciousness, a cosmic or divine root from which our consciousness has grown and with which it remains subtly connected?

The informed universe is a world of subtle but constant

interconnection, a world where everything informs—acts on and inter-acts with—everything else. This world merits deeper acquaintance; we should apprehend it with our heart as well as our brain. Chapter 9 speaks to our heart. It offers a vision that is imaginative but not imaginary: a poetic vision of a universe where nothing disappears without a trace, and where all things that exist are, and remain, intrinsically and intimately interconnected.

Science and the Akashic Field has been written to give readers interested in exploring what science can tell us about the world both the theoretical background necessary to grasp the "theory of everything" that is now within the reach of avant-garde scientists and an inkling of the vast vistas opened when this integral theory is queried about the real nature of cosmos, life, and consciousness.

Come,
sail with me on a quiet pond.
The shores are shrouded,
the surface smooth.
We are vessels on the pond
and we are one with the pond.

A fine wake spreads out behind us,
traveling throughout the misty waters.
Its subtle waves register our passage.

Your wake and mine coalesce,
they form a pattern that mirrors
your movement as well as mine.
As other vessels, who are also us,
sail the pond that is us as well,
their waves intersect with both of ours.
The pond's surface comes alive
with wave upon wave, ripple upon ripple.
They are the memory of our movement;
the traces of our being.

The waters whisper from you to me and from me to you,
and from both of us to all the others who sail the pond:

Our separateness is an illusion;
we are interconnected parts of the whole—
we are a pond with movement and memory.
Our reality is larger than you and me,
and all the vessels that sail the waters,
and all the waters on which they sail.

PART ONE

THE QUEST FOR
AN INTEGRAL THEORY
OF EVERYTHING

BACKGROUND BRIEF

WHAT ARE THEORIES OF EVERYTHING?

In the contemporary sciences, theories of everything are researched and developed by theoretical physicists. They attempt to achieve what Einstein once called "reading the mind of God." He said that if we could bring together all the laws of physical nature into a consistent set of equations, we could explain all the features of the universe on the basis of that equation; that would be tantamount to reading the mind of God.

Einstein's own attempt took the form of a unified field theory. Although he pursued this ambitious quest until his death in 1955, he did not find the simple and powerful equation that would explain all physical phenomena in a logically consistent form.

The way Einstein tried to achieve his objective was by considering all physical phenomena as the interaction of continuous fields. We now know that his failure was due to the disregard of the fields and forces that operate at the microphysical level of reality: these fields (the weak and the strong nuclear forces) are central to quantum mechanics, but not to relativity theory.

A different approach has been adopted today by the majority of theoretical physicists: they take quanta—the discontinuous aspect of physical reality—as basic. But the physical nature of quanta is reinterpreted: they are no longer discrete matter-energy particles but rather vibrating one-dimensional filaments: "strings" and "superstrings." Physicists try to link all the laws of physics as the vibration of superstrings in a higher dimensional space. They see each particle as a string that makes its own "music" together with all other particles. Cosmically, entire stars and galaxies vibrate together, as, in the final analysis, does the whole universe. The physicists' challenge is to come up with an equation that shows how one vibration relates to another, so that they can all be expressed consistently in a single super-equation. This equation would decode the encompassing music that is the vastest and most fundamental harmony of the cosmos.

At the time of writing, a string-theory-based TOE remains an ambition and a hope: nobody has come up with the super-equation that could express the harmony of the physical universe in an equation as simple and basic as Einstein's original $E = mc^2$. Yet the quest for a theory of everything is realistic. Even if an equation is found that can account for all the laws and constants of *physical* nature, a single equation is unlikely to embrace all the diverse phenomena of the world. But a single conceptual scheme could do so. And this scheme could be both simple and meaningful, as we shall see . . .

ONE

A Meaningful Worldview
for Our Time

Meaningfulness in science is an important dimension, even if it is an often neglected one. Science is not only a collection of formulas, abstract and dry, but also a source of insight into the way things are in the world. It is more than just observation, measurement, and computation; it is also a search for meaning and truth. Scientists are concerned with not only the *how* of the world—the way things work—but also *what* the things of this world are and *why* they are the way we find them.

It is indisputable, however, that many, and perhaps the majority, of physical scientists are more concerned with making their equations pan out than with the meaning they can attach to them. There are exceptions. Stephen Hawking is among those keenly interested in explicating the meaning of the latest theories, even though this is not an easy task in physics and cosmology. Shortly after the publication of his *A Brief History of Time,* a feature story appeared in the *New York Times* entitled, "Yes Professor Hawking, but what does it mean?" The question was to the point: Hawking's theory of time and the universe is complex, its meaning by no means transparent. Yet Hawking's attempts to make it so are noteworthy, and worthy of being followed up.

Evidently, the search for meaning is not confined to science. It is entirely fundamental for the human mind; it is as old as civilization. For as long as people looked at the sun, the moon, and the starry sky above, and at the seas, the rivers, the hills, and the forests below, they wondered where it all came from, where it all is going, and what it all means. In the modern world, many scientists are technical specialists, but some among them wonder as well. Theoreticians wonder more than experimentalists. They often have a deep mystical streak; Newton and Einstein are prime examples. Some scientists, the physicist David Peat among them, accept and explicitly acknowledge the challenge of finding meaning through science.

"Each of us is faced with a mystery," Peat began his book *Synchronicity*. "We are born into this universe, we grow up, work, play, fall in love, and at the ends of our lives, face death. Yet in the midst of all this activity we are constantly confronted by a series of overwhelming questions: What is the nature of the universe and what is our position in it? What does the universe mean? What is its purpose? Who are we and what is the meaning of our lives?" Science, Peat claims, attempts to answer these questions, since it has always been the province of the scientist to discover how the universe is constituted, how matter was first created, and how life began.

But other scientists do not think that contemporary science has much to do with questions of meaning. The cosmological physicist Steven Weinberg is adamant that the universe as a physical process is meaningless; the laws of physics offer no discernible purpose for human beings. "I believe there is no point that can be discovered by the methods of science," he said in an interview. "I believe that what we have found so far—an impersonal universe which is not particularly directed towards human beings—is what we are going to continue to find. And that when we find the ultimate laws of nature they will have a chilling, cold, impersonal quality about them."

This split in the scientists' view about meaning has deep cultural roots. The historian of civilization Richard Tarnas pointed out that since the dawn of the modern age, the civilization of the Western world has had two faces. One face is that of progress, the other, of fall. The more familiar face is the account of a long and heroic journey from a primitive world of dark ignorance, suffering, and limitation to the bright modern world of ever-increasing knowledge, freedom, and well-being, made possible by the sustained development of human reason and, above all, of scientific knowledge and technological skill. The other face is the story of humanity's fall and separation from the original state of oneness with nature and cosmos: while in their primordial condition humans possessed an instinctive knowledge of the sacred unity and profound interconnectedness of the world, a deep schism arose between humankind and the rest of reality with the ascendance of the rational mind. The nadir of this development is reflected in the current ecological disaster, moral disorientation, and spiritual emptiness.

Contemporary Western civilization displays both the positive and the negative faces. Its duality is reflected in the attitude scientists adopt toward the question of meaning. Some, like Weinberg, express the negative face of Western civilization. For them, meaning resides in the human mind alone: the world itself is impersonal, without purpose or intention. Finding meaning in the universe is to make the error of projecting one's own mind and personality into it. Others, like Peat, align themselves with the positive face. They insist that though the universe has been disenchanted by modern science, it is re-enchanted in light of the latest findings.

Science's disenchantment of the world has exacted a high price. When mind, consciousness, and meaning are seen as uniquely human phenomena, we humans—purposeful, valuing, feeling beings—find ourselves in a universe devoid of the very qualities we ourselves possess. We are strangers in the world in which we have come to be. Our alienation from nature opens the way to the blind exploitation of everything around us. If we arrogate all mind to ourselves, said Gregory Bateson, we will see the world as mindless and therefore as not entitled to moral or ethical consideration. "If this is your estimate of your relation to nature and you have an advanced technology," Bateson added, "your likelihood of survival will be that of a snowball in hell."

The depressive futility inherent in the negative face of Western civilization has been spelled out by the renowned philosopher Bertrand Russell: "That man is the product of causes which had no prevision of the end they were achieving," he wrote, "his hopes and fears, his loves and beliefs, are but the outcome of accidental collocations of atoms; that no fire, no heroism, no intensity of thought and feeling, can preserve an individual life beyond the grave; that all the labors of the ages, all the devotion, all the inspiration, all the noonday brightness of human genius, are destined to extinction in the vast death of the solar system, and the whole temple of man's achievement must inevitably be buried beneath the debris of a universe in ruins—all these things, if not quite beyond dispute, are yet so nearly certain, that no philosophy which rejects them can hope to stand."

But the face of progress need not be so cold, nor the face of fall so tragic. All the things that Russell mentions are not only not "beyond dispute," and not only are they not "nearly certain"; they may be the chimeras of an obsolete view of the world. At its cutting edge, the new cosmology discovers a world where the universe does not end in ruin, and the new physics, the new biology, and the new consciousness research recognize that in this world life and mind are integral elements and not accidental by-products. All these elements come together in the informed universe—a comprehensive and intensely meaningful universe, cornerstone of the unified conceptual scheme that can tie together all the diverse phenomena of the world: the *integral theory of everything*.

TWO

On Puzzles and Fables: The Next Paradigm Shift in Science

Whatever interpretation of the findings scientists may espouse, they are hard at work mapping ever more of the reality to which their observations and experiments are believed to refer. Scientists are not necessarily sophisticated philosophers, and they do not see the world in its pristine purity anymore than anyone else does. They see the world through their theories—their own conceptions about the segment of the world they investigate. However, these conceptions, unlike the ideas of philosophers and everyone else, are rigorously tested. Established theories "work": they allow scientists to make predictions based on what they observe. When they test their predictions and what they observe corresponds to what they had predicted, they maintain that their theories provide a correct account of *how* things are in that given segment of the world, *what* those things are, and *why* they are the way we actually find them. Thoroughly tested and well-developed theories about life, mind, and the universe could well be, and are even likely to be, humanly meaningful—as we shall see.*

Whether or not scientific theories are humanly meaningful, they are clearly not eternal. Occasionally even the best-established theories break

---

*The ideas and findings outlined here and in the next chapters are presented in a more detailed but also more technical form in Ervin Laszlo, *The Connectivity Hypothesis: Foundations of an Integral Science of Quantum, Cosmos, Life, and Consciousness* (Albany: State University of New York Press, 2003).

down—the predictions flowing out of them are not matched by observations. In that case the observations are said to be "anomalous"; they have no ready explanation. Strangely enough, this is the real engine of progress in science. When everything works, there can still be progress, but it is piecemeal progress at best, the refinement of the accepted theory to correspond to further observations and findings. Significant change occurs when this is not possible. Then the point is sooner or later reached when—instead of trying to stretch the established theories—scientists prefer to look for a simpler and more insightful theory. The way is open to fundamental theory innovation: to a *paradigm shift*. The shift is driven by the accumulation of observations that do not fit the accepted theories and cannot be made to fit by the simple extension of those theories. The stage may be set for a new and more adequate scientific paradigm, but that paradigm must first be discovered.

There are stringent requirements for any new paradigm. A theory based on it must enable scientists to explain all the findings covered by the previous theory, and must also explain the anomalous observations. It must integrate all the relevant facts in a simpler yet more encompassing and powerful concept. This is what Einstein did at the turn of the twentieth century when he stopped looking for solutions to the puzzling behavior of light in the framework of Newtonian physics and created instead a new concept of physical reality: the theory of relativity. As he himself said, one cannot solve a problem with the same kind of thinking that gave rise to that problem. In a surprisingly short time, the bulk of the physics community abandoned the classical physics founded by Newton and embraced Einstein's revolutionary concept in its place.

In the first decade of the twentieth century, science underwent a basic "paradigm shift." Now, in the first decade of the twenty-first century, puzzles and anomalies are accumulating again in many disciplines, and science faces another paradigm shift, very likely just as fundamental as the revolution that shifted science from the mechanistic world of Newton to the relativistic universe of Einstein.

The current paradigm shift has been brewing in the avant-garde circles of science for some time. Scientific revolutions are not instant-fit processes, with a new theory clicking into place all at once. They may be rapid, as in the case of Einstein's theory, or more protracted, as the shift from the classical Darwinian theory to a more systemic post-Darwinian conception in biology, for example. Before such revolutions are consolidated, the sciences affected by them go through a period of turbulence. Mainstream scientists defend the established theories, while maverick scientists at the cutting edge explore alternatives. The latter come up with new, sometimes radically different ideas that look at the same phenomena the mainstream scientists look at but see them differently. For a time, the alternative conceptions—initially in the form of working hypotheses—seem strange if not actually fantastic. They are something like fables, dreamt up by imaginative investigators. Yet they are not the work of untrammeled imagination. The "fables" of serious investigators are based on rigorous reasoning, bringing together what is already known about the segment of the world researched in a given discipline with what is as yet puzzling about it. And they are testable, capable of being confirmed or proved false by observation and experiment.

Investigating the anomalies that crop up in observation and experiment and coming up with the fables that could account for them make up the nuts and bolts of fundamental research in science. If the anomalies persist despite the best efforts of mainstream scientists, and if one or another of the fables advanced by maverick investigators gives a simpler and more logical explanation, a critical mass of scientists (mostly young ones) stops standing by the old paradigm. We have a paradigm shift. A concept that was until then a fable is recognized as a valid scientific theory.

There are countless examples of successful as well as of failed fables in the sciences. Confirmed fables—presently valid even if not eternally true scientific theories—include Charles Darwin's concept that all living species descended from common ancestors and Alan Guth's and Andrei Linde's hypothesis that the universe originated in a superfast "inflation" following its explosive birthing in the Big Bang. Failed fables—those that turn out not to be an exact, or at any rate the best, explanation of the pertinent phenomena—include Hans Driesch's notion that the evolution of life follows a preestablished plan in a goal-guided process called *entelechy* and Einstein's own hypothesis that an additional physical force, called the cosmological constant, keeps the universe from collapsing under the pull of gravitation. (Interestingly, as we shall see, some of these theories are being questioned again: it may be that Guth's and Linde's "inflation theory" will be replaced by the more encompassing concept of a cyclical universe, and that Einstein's cosmological constant was not mistaken after all . . .)

TWO WIDELY DISCUSSED PHYSICS FABLES

Here, by way of example, are two imaginative working hypotheses—"scientific fables"—put forward by well-respected physicists. Both have received attention well beyond the physics community, yet both are entirely mind-boggling as descriptions of the real world.

10^{100} UNIVERSES

In 1955 the physicist Hugh Everett advanced the fabulous explanation of the quantum world that was subsequently the basis for *Timeline,* one of Michael Crichton's best-selling novels. Everett's "parallel universes hypothesis" refers to a puzzling finding in quantum physics: that as long as a particle is not observed, measured, or interacted with in any way, it is in a curious state that is the superposition of all its possible states. When, however, the particle is observed, measured, or subjected to an interaction, this state of superposition becomes resolved: the particle is then in a single state only, like any "ordinary" thing. Because the state of superposition is described in a complex wave function associated with the name of Erwin Schrödinger, when the superposed state resolves it is said that the Schrödinger wave function "collapses."

The rub is that there is no way to tell which of its possible states the particle will then occupy. The particle's choice seems to be indeterminate—entirely independent of the conditions that trigger the wave function's collapse. Everett's hypothesis claims that the indeterminacy of the wave function's collapse does not reflect actual conditions in the world. There is no indeterminacy involved here: each state occupied by the particle is deterministic in itself—it simply takes place in a world of its own!

This is how the collapse would occur: When a quantum is measured, there are a number of possibilities, each of which is associated with an observer or a measuring device. We perceive only one of these possibilities in a seemingly random process of selection. But, according to Everett, the selection is not random, for it does not take place in the first place: *all* possible states of the quantum are realized every time it is measured or observed; they are just not realized in the same world. The many possible states of the quantum are realized in as many universes.

Suppose that when it is measured, a quantum such as an electron has a fifty percent probability of going up and a fifty percent probability of going down. Then we do not have just one universe in which the quantum has a 50/50 probability of going up or going down, but two parallel universes. In one of the universes the electron is actually going up and in the other it is actually going down. We also have an observer or a measuring instrument in each of these universes. The two outcomes exist simultaneously in the two universes, and so do the observers or measuring instruments.

Of course, there are not just two, but rather a vast number of possible states that a particle can occupy when its multiple superposed states resolve into a single state. Consequently, a vast number of universes must exist—perhaps of the order of 10^{100}—complete with observers and measuring instruments. Since we are not aware of any universe other than the one we observe, these universes must be separate, isolated from one another.

THE HOLOGRAPHIC UNIVERSE

The more recent "holographic universe hypothesis" advanced by particle physicists likewise boggles the mind. It claims that the entire universe is a hologram—or, at least, that it can be treated as such. Holograms, we should note, are three-dimensional representations of objects recorded with a special technique. A holographic recording consists of the pattern of interference created by two beams of light. (Currently, monochromatic lasers and semitransparent mirrors are used for this purpose.) Part of the laser light passes through the mirror and part is reflected and bounced off the object to be recorded. A photographic plate is exposed with the interference pattern created by the light beams. This is a two-dimensional pattern and it is not meaningful in itself; it is merely a jumble of lines. Nonetheless, it contains information on the contours of the object. These contours can be re-created by illuminating the plate with laser light. The patterns recorded on the photographic plate reproduce the interference pattern of the light beams, so that a visual effect appears that is identical to the 3-D image of the object. This image appears to float above and beyond the photographic plate, and it shifts according to the angle at which one views it.

The idea behind the holographic universe hypothesis is that all the information that constitutes the universe is stored on its periphery, which is a two-dimensional surface. This two-dimensional information reappears inside the universe in three dimensions. We see the universe in three dimensions even though what makes it what it is, is a two-dimensional pattern. Why is this outlandish idea the subject of intense discussion and research?

The problem the holographic universe concept addresses comes from thermodynamics. According to its solidly established second law, disorder can never decrease in any closed system. This means that disorder cannot decrease in the universe as a whole because when we take the cosmos in its totality, it is a closed system: there is no "outside" and hence nothing to which it could be open. If disorder cannot decrease, order—which can be represented as information—cannot increase. According to quantum theory, the information that creates or maintains order must be constant; it not only cannot increase, but it also cannot diminish or vanish.

But what happens to information when matter collapses into black holes? It would seem that black holes wipe out the information contained in matter. In response to this riddle, Stephen Hawking, of Cambridge University, and Jacob Bekenstein, then of Princeton University, worked out that disorder in a black hole is proportional to its surface area. Within the black hole there is a great deal more room for order and information than at its surface. In a single cubic centimeter, for example, there is room for 10^{99} Planck volumes inside, but room for only 10^{66} bits of information on the surface (a Planck volume is a space bounded by sides that measure 10^{-35} meter—an almost inconceivably small space). Now, when matter implodes into a black hole, an enormous chunk of information within the black hole seems to be wiped out. Hawking was ready to affirm that this is so, but this would fly in the face of quantum theory's assertion that in the universe, information can never be lost. The way out of this dilemma surfaced in 1993 when, working independently, Leonard Susskind, of Stanford University, and Gerard 't Hooft, of the University of Utrecht, came up with the idea that information inside the black hole is not lost if it is stored holographically on its surface.

The mathematics of holograms found unexpected application in 1998, when Juan Maldacena, then at Harvard University, tried to account for string theory under conditions of quantum gravity. Maldacena found that it is easier to deal with strings in five-dimensional spaces than in four dimensions. (We experience space in three dimensions: two planes along the surface and one up and down. A fourth dimension would be in a direction perpendicular to these, but this dimension cannot be experienced. Mathematicians can add any number of further dimensions, further and further removed from the world of experience.) The solution seemed evident: assume that the five-dimensional space inside the black hole is really a hologram of a four-dimensional pattern on its surface. One can then do the calculations in the more manageable five dimensions while dealing with a space of four dimensions.

Would this dimensional reduction work for the universe as a whole? String theorists are struggling with many extra dimensions, having discovered that three-dimensional space is not enough to accomplish their quest to come up with an equation that relates the vibrations of the various strings of the universe. Not even a four-dimensional space-time continuum will work. Initially TOEs required up to twenty dimensions to relate all vibrations together in a consistent cosmic harmony. Today scientists find that ten or eleven dimensions would suffice, provided that the vibrations occur in a higher-dimensional "hyperspace." The holographic principle—as the holographic universe hypothesis came to be known—would help: they could assume that the entire universe is a many-dimensional hologram, conserved in a smaller number of dimensions on its periphery.

The holographic principle may make string theory's calculations easier, but it makes truly fabulous assumptions about the nature of the world. (We should add that Gerard 't Hooft, one of the originators of this principle, later changed his mind about its cogency. Rather than a "principle," he said, in this context holography is actually a "problem." Perhaps, he speculated, quantum gravity could be derived from a deeper principle that does not obey quantum mechanics.)

In periods of scientific revolution, when the established paradigm is increasingly under pressure, the fables of cutting-edge researchers acquire particular importance. Some remain fabulous, but others harbor the seeds of significant scientific advance. Initially, nobody knows for sure which of the seeds will grow and bear fruit. The field is in ferment, in a state of creative chaos. This is the case today in a remarkable variety of scientific disciplines. A growing number of anomalous phenomena are coming to light in physical cosmology, in quantum physics, in evolutionary and quantum biology, and in the new field of consciousness research. They create growing uncertainties and induce open-minded scientists to look beyond the bounds of the established theories. While conservative investigators insist that the only ideas that can be considered scientific are those published in established science journals and reproduced in standard textbooks, maverick researchers look for fundamentally new concepts, including some that were considered beyond the pale of their discipline but a few years ago. As a result, the world in a growing number of disciplines is turning more and more fabulous. It is furnished with dark matter, dark energy, and multidimensional spaces in cosmology, with particles that are instantly connected throughout space-time by deeper levels of reality in quantum physics, with living matter that exhibits the coherence of quanta in biology, and with space- and time-independent transpersonal connections in consciousness research—to mention but a few of the currently advanced "fables."

Even if we do not yet know which of the fables put forward today will become accepted scientific theory tomorrow, we can already tell what *kind* of fable is likely to make it. The most promising fables have shared characteristics. In addition to being innovative and logical, they address the principal kinds of anomalies in a fundamentally new and meaningful way.

The principal kinds of anomalies today are anomalies of *coherence* and *correlation*. Coherence is a well-known phenomenon in physics: in its ordinary form, it refers to light as being composed of waves that have a constant difference in phase. Coherence means that phase relations remain constant and processes and rhythms are harmonized. Ordinary light sources are coherent over a few meters; lasers, microwaves, and other technological light sources remain coherent for considerably greater distances. But the kind of coherence discovered today is more complex and remarkable than the standard form, indicating a quasi-instant tuning together of the parts or elements of a system, whether that system is an atom, an organism, or a galaxy. All parts of a system of such coherence are so correlated that what happens to one part also happens to the other parts.

Investigators in a growing number of scientific fields are encountering this surprising form of coherence, and the correlation that underlies it. These phenomena crop up in disciplines as diverse as quantum physics, cosmology, evolutionary biology, and consciousness research and they point toward a previously unknown form and level of unity in nature. The discovery of this unity is at the core of the next paradigm shift in science. This is a remarkable development, for the new paradigm—as we shall see—offers the best-ever basis for creating the long sought but hitherto unachieved *integral theory of everything.*

THREE

A Concise Catalog
of Contemporary Puzzles

Before embarking on the search for an integral TOE, we should review the puzzles that are emerging in the pertinent fields of the sciences. We should be familiar with the unexpected and often strange findings that stress the current theories of the physical world, the living world, and the world of human consciousness, for only then can we understand the concepts that not only shed light on one or the other of these persistent domains of mystery, but also address the elements they have in common—and thus give us a new, more integral understanding of nature, mind, and universe.*

1. THE PUZZLES OF COSMOLOGY

Cosmology, a branch of the astronomical sciences, is in turbulence. The deeper the new high-powered instruments probe the far reaches of the universe, the more mysteries they uncover. For the most part, these mysteries have a common element: they exhibit a staggering coherence throughout the reaches of space and time.

---

*This catalog offers a preliminary overview; a fuller account is given in chapter 5.

THE SURPRISING WORLD OF
THE NEW COSMOLOGY

THE PRINCIPAL LANDMARK: THE COHERENTLY
STRUCTURED AND EVOLVING COSMOS

The universe is far more complex and coherent than anyone other than poets and mystics have dared to imagine. A number of puzzling observations have cropped up:

- *The "flatness" of the universe:* in the absence of matter, space-time turns out to be "flat" or "Euclidean" (the kind of space where the shortest distance between two points is a straight line), rather than curved (where the shortest distance between any two points is a curve). This, however, means that the "Big Bang" that gave rise to our universe was staggeringly finely tuned, for if it had produced just one-billionth more or one-billionth less matter than it did, space-time would be curved even in the absence of matter.

- *The "missing mass" of the universe:* there is more gravitational pull in the cosmos than visible matter can account for—yet only matter is believed to have mass and thus to exert the force of gravitation. Even when cosmologists allow for a variety of "dark" (optically invisible) matter, there is still a great chunk of matter (and hence mass) missing.

- *The accelerating expansion of the cosmos:* distant galaxies pick up speed as they move away from each other—yet they should be slowing down as gravitation brakes the force of the Big Bang that blew them apart.

- *The coherence of some cosmic ratios:* the mass of elementary particles, the number of particles, and the forces that exist between them are all mysteriously adjusted to favor certain ratios that recur again and again.

- *The "horizon problem"*: the galaxies and other macrostructures of the universe evolve almost uniformly in all directions from Earth, even across distances so great that the structures could not have been connected by light, and hence could not have been correlated by signals carried by light (according to relativity theory, no signal can travel faster than light).
- *The fine-tuning of the universal constants*: the key parameters of the universe are amazingly fine tuned to produce not just recurring harmonic ratios, but also the—otherwise extremely improbable—conditions under which life can emerge and evolve in the cosmos.

According to the standard model of cosmic evolution, the universe originated in the Big Bang, twelve to fifteen billion years ago (the latest satellite-based observations, made from the far side of the moon, confirm that the universe is indeed about 13.7 billion years old). The Big Bang was an explosive instability in the "pre-space" of the universe, a fluctuating sea of virtual energies known by the misleading term *vacuum*. A region of this vacuum—which was, and is, far from a real vacuum, that is, empty space—exploded, creating a fireball of staggering heat and density. In the first milliseconds it synthesized all the matter that now populates cosmic space. The particle-antiparticle pairs that emerged collided with and annihilated each other, and the one billionth of the originally created particles that survived (the tiny excess of particles over antiparticles) made up the material content of this universe. After about 200,000 years, the particles decoupled from the radiation field of the primordial fireball, space became transparent, and clumps of matter established themselves as distinct elements of the cosmos. Matter in these clumps condensed under gravitational attraction: the first stars appeared about 200 million years after the Big Bang. In the space of one billion years, the first galaxies were formed.

Until quite recently, the scenario of cosmic evolution seemed well established. Detailed measurements of the cosmic microwave background radiation—the presumed remnant of the Big Bang—testify that its variations derive from minute fluctuations within the cosmic fireball when our universe was less than one trillionth of a second "young" and are not distortions caused by radiation from stellar bodies.

However, the standard cosmology of the Big Bang is not as established now as it was a few years ago. There is no reasonable explanation in "BB theory" for the observed flatness of the universe; for the missing mass in it; for the accelerating expansion of the galaxies; for the coherence of some basic cosmic ratios; and for the "horizon problem," the uniformity of the macrostructures throughout cosmic space. The problem known as the "tuning of the constant" is particularly vexing. The three dozen or more physical parameters of the universe are so finely tuned that together they create the highly improbable conditions under which life can emerge on Earth (and presumably on other suitable planetary surfaces) and then evolve to progressively higher levels of complexity. These are all puzzles of coherence, and they raise the possibility that this universe did not arise in the context of a random fluctuation of the underlying quantum vacuum. Instead, it may have been born in the womb of a prior "meta-universe": a *Metaverse*. (The term *meta* comes from classical Greek, signifying "behind" or "beyond," in this case meaning a vaster, more fundamental universe that is behind or beyond the universe we observe and inhabit.)

The existence of a vaster, perhaps infinite universe is underscored by the astonishing finding that no matter how far and wide high-powered telescopes range in the universe, they find galaxy after galaxy—even in "black regions" of the sky where no galaxies or stars of any kind were believed to exist. This picture is a far cry from the concept that reigned in astronomy but a hundred years ago. At that time, and until the 1920s, it was thought that the Milky Way was all there is to the universe: where the Milky Way ends, space itself ends. Not only do we know today that the Milky Way—"our galaxy"—is but one among billions of other galaxies in "our universe," but we are also beginning to recognize that the boundaries of "our universe" are not the boundaries of "*the* universe." The cosmos may be infinite in time, and perhaps also in space—it is vaster by several magnitudes than any cosmologist would have dared to dream just a few decades ago.

Today a number of physical cosmologies offer quantitatively elaborated accounts of how the universe we inhabit could have arisen in the framework of a Metaverse. The promise of such cosmologies is that they may overcome the puzzles of coherence in this universe, including the mind-blowing serendipity that it is so improbably finely tuned that we can be here to ask questions about it. This has no credible explanation in a one-shot, single-cycle universe, for there the pre-space fluctuations that set the parameters of the emerging universe must have been randomly selected: there was "nothing there" that could have biased the serendipity of this selection. Yet a random selection from among all the possible fluctuations in the chaos of a turbulent pre-space is astronomically unlikely to have led to a universe where living organisms and other complex and coherent phenomena could arise and evolve!

The fluctuations that led to our amazingly coherent universe may not have been selected at random. Traces of prior universes could have been present in the pre-space from which our universe arose. They could have reduced the range of the fluctuations that affected the explosion that created our universe, fine-tuning the fluctuations to those that lead to a universe that can give rise to complex systems, such as those required for life. In this way the Metaverse could have informed the birth and evolution of our universe, much as the genetic code of our parents informed the conception and growth of the embryo that grew into what we are today.

The staggering coherence of our universe tells us that all its stars and galaxies are interconnected in some way. And the astonishing fine-tuning of the physical laws and constants of our universe suggests that at its birth our universe may have been connected with prior universes in a vaster, perhaps infinite Metaverse.

Do we come across here the footprint of a cosmic "Akashic Field" that conveyed the trace of a precursor universe to the birth of our universe—and has been connecting and correlating the stars and galaxies of this universe ever since?

2. THE PUZZLES OF QUANTUM PHYSICS

In the course of the twentieth century, quantum physics—the physics of the ultrasmall domain of physical reality—became strange beyond imagination. The discoveries show that the smallest identifiable units of matter, force, and light are actually made up of energy, but not a continuous flow of energy: they always come in distinct packets known as quanta. These energy packets are not material, although they can have matterlike properties such as mass, gravitation, and inertia. They seem like objects, but they are not ordinary, commonsense objects: they are both corpuscles and waves. When one of their properties is measured, the others become unavailable to measurement and observation. And they are instantly and nonenergetically "entangled" with each other no matter how far apart they may be.

At the quantum level, reality is strange and it is nonlocal: the whole universe is a network of time- and space-transcending interconnection.

THE WEIRD WORLD OF THE QUANTUM

THE PRINCIPAL LANDMARK: THE ENTANGLED PARTICLE

- In their pristine state, quanta are not just in one place at one time: each single quantum is both "here" and "there"—and in a sense it is everywhere in space and time.

- Until they are observed or measured, quanta have no definite characteristics but instead exist simultaneously in several states at the same time. These states are not "real" but "potential"—they are the states the quanta can assume when they are observed or measured. (It is as if the observer, or the

---

measuring instrument, fishes the quanta out of a sea of possibilities. When a quantum is pulled out of that sea, it becomes a real rather than a mere virtual beast—but one can never know in advance just which of the various real beasts it *could* become it actually *will* become. It appears to choose its real states on its own.)

- Even when the quantum is in a set of real states, it does not allow us to observe and measure all of these states at the same time: when we measure one of its states (for example, position or energy), another becomes blurred (such as its speed of motion or the time of its observation).

- Quanta are highly sociable: once they are in the same state, they remain linked no matter how far they travel from each other. When one of the formerly connected quanta is subjected to an interaction (that is, when it is observed or measured), it chooses its own state—and its twin also chooses its own state, but not freely: it chooses it according to the choice of the first twin. It always chooses a complementary state, never the same one.

- Within a complex system (such as the whole setup of an experiment), quanta exhibit just as sociable behaviors. If we measure one of the quanta in the system, the others become "real" (that is, similar to a commonsense object) as well. Even more remarkably, if we create an experimental situation where a given quantum can be individually measured, all the other quanta become "real" even if the experiment is *not* carried out . . .

Classical mechanics, the physics of Isaac Newton, conveyed a comprehensible concept of physical reality. Newton's *Philosophiae Naturalis Principia Mathematica,* published in 1687, demonstrated with geometrical precision that material bodies move according to mathematically expressible rules on Earth, while planets rotate in accordance with Kepler's laws in the heavens. The motion of all things is rigorously determined by the conditions under which it is initiated, just as the motion of a pendulum is determined by its length and its initial displacement and that of a projectile by its launch angle and acceleration. With mathematical certainty Newton predicted the position of the planets, the motion of pendulums, the path of projectiles, and the motion of the "mass points" that in his physics are the ultimate building blocks of the universe.

Somewhat over a hundred years ago, the mechanistic, predictable world of Newton ran into trouble. With the splitting of the atom in the late nineteenth century and of the atomic nucleus in the early twentieth, more had been fragmented than a physical entity. The very foundation of natural science was shaken: the experiments of early-twentieth-century physics demolished the prevailing view that all of reality is built of blocks that are themselves not further divisible. Yet physicists could not put any comparably commonsensical concept in its place. The very notion of "matter" became problematic. The subatomic particles that emerged when atoms and atomic nuclei were fissioned did not behave like conventional solids: they had a mysterious interconnection known as "nonlocality," and a dual nature consisting of wavelike as well as corpuscle-like properties. In addition, the famous "EPR" experiment (the experiment originally suggested by Albert Einstein together with colleagues Boris Podolski and Nathan Rosen) demonstrated that particles that at one time shared the same system of coordinates remain instantly and enduringly correlated. Such correlation extends to entire atoms: current "teleportation" experiments show that when one of a pair of correlated atoms is further correlated with a third atom, the quantum state of the third is instantly transferred ("beamed") to the other of the initially correlated pair—no matter how far away that atom may be. . . .

The remarkable fact emerging from this sea of quantum mystery is that particles and atoms are not individual beasts. They are sociable entities, and under certain conditions they are so thoroughly "entangled" with each other that they are not just here or there, but in all pertinent places at the same time. Their nonlocality respects neither time nor space: it exists whether the distance that separates the particles and the atoms is measured in millimeters or in light-years, and whether the time that separates them consists of seconds or of millions of years.

> *Could the nonlocality of the most basic elements of the universe be due to a fundamental field that records the state of particles and atoms and conveys this information to particles and atoms in corresponding states? Could it be that an Akashic Field is active not only at the cosmological scale, but also at the ultra-small scale of physical reality?*

3. THE PUZZLES OF BIOLOGY

The superlarge as well as the ultrasmall domains of physical reality turn out to be amazingly correlated and coherent. But the world in its every-day dimension is more reasonable. Here things occupy but one state at a time and are either here or there and not in both places simultaneously. This, at any rate, is the commonsense assumption, but in regard to living beings, it turns out not to be true. This is surprising, for the living organism is made up of cells, which are made up of molecules, which in turn are made up of atoms, made up of particles. And even if particles themselves are weird, the whole made up of them should be a classical, commonsense object: one would expect that quantum inde-terminacies would be canceled out at the macroscale.

But in the living world, macroscale objects are not classical—or not entirely so. Instant, multidimensional correlations are coming to light between the parts of a living organism, and even between organisms and environments. Cutting-edge research in quantum biology finds that atoms and molecules in the organism, and even entire organisms and their environments, are nearly as "entangled" as microparticles that originate in the same quantum state.

THE UNEXPECTED WORLD OF POST-DARWINIAN BIOLOGY

THE PRINCIPAL LANDMARK: THE SUPER-COHERENT ORGANISM

- The living organism is extraordinarily coherent: all its parts are multidimensionally, dynamically, and almost instantly correlated with all other parts. What happens to one cell or organ also happens in some way to all other cells and organs—a correlation that recalls (and in fact suggests) the kind of "entanglement" that characterizes the behavior of quanta in the microdomain.

- The organism is also coherent with the world around it: what happens in the external milieu of the organism is reflected in some ways in its internal milieu. Thanks to this coherence, the organism can evolve in tune with its environment. The genetic makeup of even a simple organism is so complex, and its "fit" to the milieu so delicate, that in the absence of such "inside-outside tuning," living species could not mutate into a viable form before being eliminated by natural selection. That our world is not populated solely by the simplest of organisms, such as bacteria and blue-green algae, is due in the last analysis to the kind of "entanglement" that exists among genes, organisms, organic species, and their niches within the biosphere.

That the living organism is coherent as a whole is not surprising—what is surprising is the degree and form of its coherence. The organism's coherence goes beyond the coherence of a biochemical system; in some respects it attains the coherence of a *quantum* system.

Evidently, if living organisms are not to succumb to the constraints of the physical world, their component parts and organs must be precisely yet flexibly correlated with each other. Without such correlation, physical processes would soon break down the organization of the living state, bringing it closer to the inert state of thermal and chemical equilibrium in which life as we know it is impossible. Near-equilibrium systems are largely inert, incapable of sustaining processes such as metabolism and reproduction, essential to the living state. An organism is in thermodynamic equilibrium only when it is dead. As long as it is living, it is in a state of *dynamic* equilibrium in which it stores energy and information and has them available to drive and direct its vital functions.

On closer analysis it turns out that dynamic equilibrium requires a very high degree of coherence: it calls for instantaneous long-range correlations throughout the system. Simple collisions among neighboring molecules—mere billiard-ball push-impact relations among them—must be complemented by a network of instant communication that correlates all parts of the living system, even those that are distant from one another. Rare molecules, for example, are seldom contiguous, yet they find each other throughout the organism. There would not be sufficient time for this to occur by a random process of jiggling and mixing; the molecules need to locate and respond to each other specifically, even if they are distant. It is difficult to see how this could be achieved by mechanical or chemical connections among the organism's parts, even if correlated by a nervous system that reads biochemical signals from genes through DNA, RNA, proteins, enzymes, and neural transmitters and activators.

In a complex organism the challenge of order is gigantic. The human body consists of some million billion cells, far more than stars in the Milky Way galaxy. Of this cell population, 600 billion are dying and the same number are regenerating every day—over 10 million cells per second. The average skin cell lives only for about two weeks; bone cells are renewed every three months. Every ninety seconds millions of antibodies are synthesized, each from about twelve hundred amino acids, and every hour 200 million erythrocytes are regenerated. There is no substance in the body that is constant, though heart and brain cells endure longer than most. And the substances that coexist at a given time produce thousands of biochemical reactions in the body each and every second.

The level of coherence exhibited by organisms suggests that quantum-type processes take place in them. For example, organisms respond to extremely low frequency electromagnetic radiation, and to magnetic fields so weak that only the most sophisticated instruments can register them. But radiation below molecular dimensions could not affect molecular assemblies unless a large number of molecules were super-coherently linked among themselves. Such linkages could come about only if quantum processes complement the organism's biochemical processes. The living organism, it appears, is in some respects a "macroscopic quantum system."

Correlation within the organism embraces the set of the organism's genes, the so-called genome. This is an anomaly for mainstream biology. According to classical Darwinism, the genome should be insulated from the vicissitudes that befall the rest of the organism. There is to be a full and complete separation of the *germ line* (the genetic information handed down from parent to offspring) from the *soma* (the organism that expresses the genetic information). Darwinists claim that in the course of successive generations in the life of a species, the germ line varies randomly, unaffected by influences acting on the soma. Evolution proceeds by a selection from among the randomly created genetic variants according to the "fit" of the soma (the resulting organism) to its particular environment. Thus biological evolution is the product of a twofold chance: the chance variation of the genome and the chance fit of the resulting mutants to their environment. To cite the metaphor made popular by the Oxford biologist Richard Dawkins, evolution occurs through trial and error: the work of a blind watchmaker.

However, the classical Darwinian tenet regarding the isolation of the genome is not correct. It has been proved false indirectly, through statistical probability, and empirically, by way of laboratory experiments. Genome, organism, and environment form an integrated system where functionally autonomous parts are so correlated that the organism can survive, and can produce offspring that prove viable under conditions that would have been fatal to the parent.

The connection between genes and environments is demonstrated in laboratory experiments. Gene–environment connection can be conveyed even by mechanical means. The cell biologist A. Maniotis described an experiment where a mechanical force impressed on an external cellular membrane was transmitted to the cell nucleus. This produced a mutation almost instantly. The experimentalist Michael Lieber went further. His work demonstrated that mechanical force acting on the outer membrane of cells is but one variety of interaction that results in a genetic rearrangement: any stress coming from the environment, mechanical or not, triggers a global "hypermutation." The genome is dynamic and highly adaptive. When challenged it creates a complex and practically instant series of rearrangements, producing even in-themselves-unnecessary steps if they facilitate the necessary steps.

The recently discovered "adaptive response" of the genome is also evident when electromagnetic or radioactive fields irradiate the organism: this, too, has a direct effect on the structure of its genes. In many cases the new arrangement shows up in the offspring. Experiments in Japan and the United States show that rats develop diabetes when a drug administered in the laboratory damages the insulin-producing cells of their pancreas. These diabetic rats produce offspring in which diabetes arises spontaneously! It appears that the alteration of the rats' body cells produces a rearrangement of their genes.

Even more striking are experiments in which particular genes of a strain of bacterium are rendered defective—for example, genes that enable bacteria to metabolize lactose. When these bacteria are fed a pure milk diet, some among them mutate back precisely those of their genes that enable them to metabolize it again. Given the complexity of the genome even of humble bacteria, this response is astronomically unlikely to occur purely by chance.

Exposure to chemicals also produces adaptive mutation. When

plants and insects are subjected to toxic substances, they often mutate their gene pool in precisely such a way that detoxifies the poisons and creates resistance to them.

The German theoretician Marco Bischof summed up the key insight currently emerging at the frontiers of the life sciences. "Quantum mechanics has established the primacy of the inseparable whole. For this reason," he said (and the emphasis is his), "the basis of the new biophysics must be the insight into the fundamental interconnectedness *within* the organism as well as *between* organisms, and that of the organism *with the environment."*

> *Could a field, sometimes called "biofield," instantly and continuously coordinate the myriad interactions of the organism's myriad molecules, genes, and cells, and correlate entire organisms and species with their environment? Could it be that the Akashic Field we have encountered in microphysics and in cosmology is also active in the domains of life—that it interconnects organisms and ecologies, much as it interconnects quanta at the ultrasmall scale of reality and the universe at the superlarge scale?*

4. THE PUZZLES OF CONSCIOUSNESS RESEARCH

Consciousness is the most intimately and immediately known fact of our experience. It accompanies us from birth, presumably until death. It is unique, and seems to belong uniquely to each of us. Yet "my" consciousness may not be solely and uniquely mine. The connections that bind "my" consciousness to the consciousness of others, well known to traditional—so-called primitive, but in fact in many respects highly sophisticated—peoples, are rediscovered today in controlled experiments with thought and image transference, and the effect of the mind of one individual on the body of another.

THE TRANSPERSONAL WORLD OF HUMAN CONSCIOUSNESS

THE PRINCIPAL LANDMARK: THE CONNECTEDNESS OF THE HUMAN MIND

- Native tribes seem able to communicate beyond the range of eye and ear. As shown by the customs, buildings, and artifacts of diverse peoples who lived on different points of the globe, and may have lived at different times, entire cultures appear to have shared information among themselves, even though they were not in any known form of contact with each other.

- In the laboratory also, modern people display a capacity for spontaneous transference of impressions and images, especially when they are emotionally close to each other.

- Some images and ideas—universal symbols and archetypes—occur and recur in the culture of all civilizations, modern and ancient, whether or not their people have known each other or have even known of each other's existence.

- The mind of one person appears able to act on the brain and body of another. This faculty, known to traditional peoples, is verified today in controlled experiments and forms the basis of a new branch of medicine known as *telesomatic* or *nonlocal* medicine.

Current findings at the farther reaches of human consciousness recall Einstein's pronouncement half a century ago. "A human being" he said, "is part of the whole, called by us 'universe,' a part limited in time and space. He experiences his thoughts and feelings as something separate from the rest—a kind of optical delusion of his consciousness. This delusion is a sort of prison for us, restricting us to our personal decisions and to affection for a few persons nearest us." While in the conservative view human communication and interaction is limited to our sensory channels (everything that is in the mind, it is said, must first have been in the eye or ear), leading psychologists, psychiatrists, and consciousness researchers are rediscovering what Einstein realized and ancient cultures have always known: that we are linked by more subtle and encompassing connections as well. In current scientific literature these connections are called *transpersonal.*

Traditional cultures did not regard transpersonal connections with distant peoples, tribes, or cultures as illusion, but modern societies do. The modern mind is not ready to accept anything as real that is not "manifest"—not literally "ready to hand" (*manus* being Latin for "hand"). Consequently, transpersonal connections are viewed as paranormal and admitted only under exceptional conditions.

One of the exceptions is "twin pain"—when one of a pair of identical twins senses the pain or trauma of the other. This phenomenon is well documented. Guy Playfair, who wrote the book *Twin Telepathy,* noted that about thirty percent of twins experience telepathic interconnection. He cites a 1997 television program where the production team tested four pairs of identical twins. The brain waves, blood pressure, and galvanic skin response of the four pairs of twins were rigorously monitored. One of the unsuspecting twins in each pair was subjected to a loud alarm fitted to the back of the chair in which he or she was sitting. In three of the four pairs, the other twin registered the resulting shock, even through he or she was closeted some distance away in a separate and soundproof room. The successful pairs were used for the show that went live on the air, and they again showed the telepathic information transmission, although the receiving twin could not give an account of what it was that the other twin had experienced. The technical supervisor of the show concluded that the twins "certainly picked up something from somewhere."

Identical twins are only the top of the tree of bonded pairs. Some form of telepathy has been observed among all people who share a deep bond, such as mothers and children, lovers, long-term couples, even close friends. In these cases all but the most conservative psychologists are forced to recognize the existence of some transpersonal contact. But only exceptionally broad-minded psychologists admit that transpersonal contact includes the ability to transmit thoughts and images, and that it is given to many and perhaps all people. Yet this is the finding of recent experiments. The telepathic powers of people—their ability to carry out various forms of thought and image transference—is not just wishful thinking or a misreading of the results. A whole spectrum of experimental protocols has been developed, ranging from the noise-reduction procedure known as the Ganzfeld technique to the rigorous "distant mental influence on living systems" (DMILS) method. Explanations in terms of hidden sensory cues, machine bias, cheating by subjects, and experimenter incompetence or error have all been considered, but were found unable to account for a number of statistically significant results. It appears that almost all people possess "paranormal" abilities.

Not only can people communicate with the minds of other people, but they can also interact with other people's *bodies*. Reliable evidence is becoming available that the conscious mind of one person can produce repeatable and measurable effects on the body of another. These effects, in turn, are known as *telesomatic*.

Telesomatic effects were known to so-called primitive peoples: anthropologists call them "sympathetic magic." Shamans, witch doctors, and those who practice such magic (voodoo, for example) do not act on the person they target, but rather on an effigy of that person, such as a doll. This practice is widespread among traditional peoples. In his famous study *The Golden Bough*, Sir James Frazer noted that Native American shamans would draw the figure of a person in sand, ashes, or clay and then prick it with a sharp stick or do it some other injury. The corresponding injury was said to be inflicted on the person the figure represented. Observers found that the targeted person often fell ill, became lethargic, and sometimes even died.

There are positive variants of sympathetic magic today that are increasingly widely known and practiced. One variant is the kind of alternative medicine known as spiritual healing. The healer acts on the organism of his or her patient by "spiritual" means—that is, by sending a healing force or healing information. Healer and patient can be directly face to face, or miles apart; distance does not seem to affect the outcome. The effectiveness of this kind of healing may be surprising, but it is well documented. Renowned physician Larry Dossey calls the corresponding medical practice "Era III nonlocal medicine," suggesting that it is the successor to Era I biochemical medicine, and Era II psychosomatic medicine.

Another form of positively oriented sympathetic magic is healing by intercessory prayer. The effectiveness of prayer has been known to religious people and communities for hundreds and indeed thousands of years. But the credit for documenting it in a controlled experiment is due to the heart specialist Randolph Byrd. He undertook a ten-month computer-assisted study of the medical histories of patients at the coronary care unit at San Francisco General Hospital. As reported in the *Southern Medical Journal* in 1988, Byrd formed a group of experimenters made up of ordinary people whose only common characteristic was a habit of regular prayer in Catholic or Protestant congregations around the country. The selected people were asked to pray for the recovery of a group of 192 patients; another set of 210 patients, for whom nobody prayed, made up the control group. Neither the patients, nor the nurses and doctors knew which patients belonged to which group. The people who were to pray were given the names of the patients and some information about their heart condition. As each person could pray for several patients, all patients had between five and seven people praying for them. The results were significant. The prayed-for group was five times less likely than the control group to require antibiotics (three compared to sixteen patients); it was three times less likely to develop pulmonary edema (six versus eighteen patients); none in the prayed-for group required endotracheal incubation (while twelve patients in the control group did); and fewer patients died in the former than in the latter group (though this particular result was statistically not significant). It did not matter how close or far the patients were to those who prayed for them, nor did the manner of praying make any difference. Only the fact of concentrated and repeated prayer was a factor, without regard to whom the prayer was addressed and where the prayers took place.

Intercessory prayer and spiritual healing, together with other mind- and intention-based experiments and practices, yield impressive evidence regarding the effectiveness of telepathic and telesomatic information- and energy-transmission. The pertinent practices produce real and measurable effects on people, and they are more and more widespread. But mainstream science has no explanation for them.

Could it be that our consciousness is linked with other consciousnesses through an interconnecting Akashic Field, much as galaxies are linked in the cosmos, quanta in the microworld, and organisms in the world of the living? And could this be the same field we have encountered before, manifesting itself in the realm of mind, in addition to the realms of nature?

FOUR

Searching for the Memory
of the Universe

Our review of the puzzles encountered in contemporary science has set the stage for the quest to which this book is dedicated: to achieve a scientifically founded integral theory of everything. We have gained an important insight. We have found that in order to account for a growing number of things and processes that are undoubtedly real and are likely to be fundamental, we must recognize that there is more to the world than the current paradigm of science allows for.

Let us look again at the principal findings:

- The universe as a whole manifests fine-tuned correlations that defy commonsense explanation.
- Astonishingly close correlations exist on the level of the quantum: every particle that has ever occupied the same quantum state as another particle remains correlated with it in a mysterious, nonenergetic way.
- Post-Darwinian evolutionary theory and quantum biology discover similarly puzzling correlations within the organism and between the organism and its milieu.
- The correlations that come to light in the farther reaches of consciousness research are just as strange: they are in the form of "transpersonal connections" between the consciousness of one person and the mind and body of another.

When we review these puzzles of connection and correlation, we come to a remarkable conclusion. The networks of connection that make for a coherently evolving cosmos, for the entanglement of quanta, for the instant connection between organisms and environments and between the consciousnesses of different and even far removed human beings, have one and the same explanation. There is not only matter and energy in the universe, but also a more subtle yet real element: information in the form of active and effective "in-formation." In-formation of this kind connects all things in space and time—indeed it connects all things *through* space and time. As a number of cutting-edge scientists, among them Nikola Tesla, then David Bohm, and more recently Harold Puthoff, surmised, *interactions in the domains of nature as well as of mind are mediated by a fundamental information field at the heart of the universe.*

ON THE TRACK OF NATURE'S INFORMATION FIELD

In the beginning of the twentieth century, the much neglected—but now more and more rediscovered—genius Nikola Tesla, the father of modern communication technologies, spoke of an "original medium" that fills space and compared it to Akasha, the light-carrying ether. In his unpublished 1907 paper "Man's greatest achievement," he wrote that this original medium, a kind of force field, becomes matter when prana, cosmic energy, acts on it, and when the action ceases, matter vanishes and returns to Akasha. Since this medium fills all of space, everything that takes place in space can be referred to it. The curvature of space, said Tesla, which was put forward at the time by Einstein, is not the answer.

However, by the end of the first decade of the twentieth century, physicists adopted Einstein's mathematically elaborated four-dimensional curved space-time and, with the exception of a few maverick theoreticians, refused to consider any concept of a space-filling ether, medium, or force field. Tesla's insight fell into disrepute, and then into oblivion. Today it is revived. Bohm, Puthoff, and a small but growing group of scientists are rediscovering the role of information in nature, and locating nature's information field in the quantum vacuum, the much discussed if as yet imperfectly understood energy sea that fills cosmic space.

BACKGROUND BRIEF

THE QUANTUM VACUUM

The concept of space-time as an energy-filled substratum of the universe emerged in the course of the twentieth century. At the beginning of that century, space was already believed to be filled with an invisible energy field—the luminiferous ether—that produces friction when bodies move through it and thus slows their motion. But when such friction failed to materialize in the famous Michelson-Morley experiments, the ether was removed from the physicists' world picture. The absolute vacuum—space that is truly empty when not occupied by matter—took its place.

However, the cosmic vacuum turned out to be far from empty space. In the "grand unified theories" (GUTs) developed in the second half of the twentieth century, the concept of the vacuum transformed from empty space into the medium that carries the zero-point field, or ZPF. (The name derives from the fact that in this field energies prove to be present even when all classical forms of energy vanish: at the absolute zero of temperature.) In subsequent unified theories, the roots of all of nature's fields and forces were ascribed to the mysterious energy sea known as the "unified vacuum."

More and more interactions have come to light between this fundamental field and the observed things and processes of the physical world. In the 1960s Paul Dirac showed that fluctuations in fermion fields (fields of matter particles) produce a polarization of the ZPF of the vacuum, whereby the vacuum in turn affects the particles' mass, charge, spin, or angular momentum.

At around the same time, Andrei Sakharov proposed that relativistic phenomena (the slowing down of clocks and the shrinking of yardsticks near the speed of light) are the result of effects induced in the vacuum due to the shielding of the zero-point field by charged particles. This is a revolutionary idea, since in this concept the vacuum is more than relativity theory's four-dimensional continuum: it is not just the geometry of space-time, but a real physical field producing real physical effects.

The physical interpretation of the vacuum in terms of the zero-point field was reinforced in the 1970s, when Paul Davis and William Unruh put forward a hypothesis that differentiates between uniform and accelerated motion in the zero-point field. Uniform motion would not disturb the ZPF, leaving it isotropic (the same in all directions), whereas accelerated motion would produce a thermal radiation that breaks open the field's all-directional symmetry. During the 1990s, numerous explorations were undertaken on this premise, going well beyond the already "classical" Casimir force and Lamb shift.

The Casimir force is well known. Between two closely placed metal plates, some wavelengths of the vacuum's energies are excluded, and this reduces the vacuum's energy density with respect to the vacuum energies on the outer side of the plates. The disequilibrium creates a pressure—this is the "Casimir force"—that pushes the plates inward and together. The Lamb shift, another thoroughly investigated vacuum effect, consists of the frequency shift exhibited by the photons that are emitted when electrons around the nucleus of an atom leap from one energy state to another. The shift is due to the photon exchanging energy with the ZPF.

Further effects have been found. Harold Puthoff, Bernhard Haisch, and collaborators produced a sophisticated theory according to which the inertial force, the gravitational force, and even mass are consequences of the interaction of charged particles with the ZPF. Puthoff also noted that electrons orbiting atomic nuclei constantly radiate energy, so that they would move progressively closer to the given nucleus were it not that the quantum of energy they absorb from the vacuum offsets the energy lost due to their orbital motion.

Even the stability of our planet in its orbit around the Sun derives from vacuum-energy inputs. As Earth pursues its orbital path, it loses momentum; given a constant loss of momentum, the gravitational field of the Sun—in the absence of an influx of energy from the ZPF—would overcome the centrifugal force that pushes Earth around its orbit and Earth would spiral into the Sun. This means that in addition to inertia, gravity, and mass, the very stability of both atoms and solar systems is due to interaction with the zero-point field of the vacuum.

Although much remains to be discovered about the quantum vacuum, it is already clear that it is a superdense cosmic medium. It carries light, and all the universal forces of nature. Pressure waves may propagate through it, traversing the universe from one end to the other. This is the finding of the German mathematical physicist Hartmut Mueller, who claims that the observed dimension of all entities, from atoms to galaxies, is determined by interaction with density-pressure waves propagating in the vacuum. According to his "global scaling theory," the universe is dimensionally limited: on the lower end of the dimensional horizons, matter density is the greatest, and on the upper end it is the least. This is due to vacuum-based pressure waves. Because the universe is finite, at the critical dimension points the waves superpose and create enduring standing waves. These waves determine physical interactions by setting the value of the gravitational, the electromagnetic, and the strong and weak nuclear forces. By means of resonance they amplify some vibrations and repress others; they are thus responsible for the distribution of matter throughout the cosmos. All processes have an inner rhythm according to their resonance with the vacuum's standing waves. Mueller concludes that the vacuum is a cosmic ultraweak background that acts as a morphogenetic field.

Recent findings confirm the presence of pressure waves in the vacuum. Astronomers in NASA's Chandra X-ray Observatory found a wave generated by the supermassive black hole in the Perseus cluster of galaxies, some 250 million light-years from Earth. This vacuum-pressure wave translates into the musical note B flat. This is a real note that has been traveling through the vacuum for the past 2.5 billion years. Our ears cannot perceive it: its frequency is fifty-seven octaves below middle C—more than a million billion times deeper than the limits of human hearing.

A field that transports light (that is, waves of photons) and density-pressure waves, and replenishes the energy lost by atoms and solar systems, is not an abstract theoretical entity. No wonder that more and more physicists speak of the quantum vacuum as a physically real cosmic *plenum*.

The quantum vacuum, it appears, transports light, energy, pressure, and sound. Could it have a further property by means of which it correlates separate and possibly distant events? Could it create the correlations that make for the amazing coherence of the quantum, of the organism, of consciousness—and of the whole universe? The vacuum could indeed have such a property. It could be not just a superdense sea of *energy*, but also a sea of *information*.

The possibility that the quantum vacuum could convey information has been raised by a number of avant-garde investigators. For example, Harold Puthoff remarked, ". . . on the cosmological scale a grand hand-in-glove equilibrium exists between the ever-agitated motion of matter on the quantum level and the surrounding zero-point energy field. One consequence of this is that we are literally, physically, 'in touch' with the rest of the cosmos as we share with remote parts of the universe fluctuating zero-point fields of even cosmological dimensions." And, Puthoff added, "[w]ho is to say whether, for example, modulation of such fields might not carry meaningful information as in the popular concept of 'the Force'?" The experiences of the *Apollo* astronaut Edgar Mitchell while in space led him to the same conclusion. According to Mitchell, information is part of the very substance of the universe. It is one part of a "dyad" of which the other part is energy. Information is present everywhere, and has been present since the birth of the universe. The quantum vacuum, Mitchell said, is the holographic information mechanism that records the historical experience of matter.

HOW THE QUANTUM VACUUM GENERATES, CONSERVES, AND CONVEYS INFORMATION

How could the quantum vacuum convey the "historical experience of matter"? This is a fundamental question for contemporary physics and possibly the key to the emerging paradigm of all sciences. There are innovative theories that promise an exciting and scientifically valid answer.

A particularly promising theory is the work of the Russian physicists G. I. Shipov, A. E. Akimov, and coworkers, further elaborated by scientists in America as well as Europe. Their "torsion-wave" theory shows how the vacuum can link physical events throughout space-time. According to the Russian physicists, torsion waves link the universe at a group speed of the order of 10^9 c—one billion times the speed of light!

Torsion-wave linking may involve more than the known forms of energy: it may also involve *information*. It is standard knowledge that particles that have a quantum property known as "spin" also have a magnetic effect: they possess a specific magnetic momentum. The magnetic impulse is registered in the vacuum in the form of minute vortices. Like vortices in water, vacuum-based vortices have a nucleus around which circle other elements—H_2O molecules in the case of water, virtual bosons (vacuum-based force particles) in the case of the zero-point field. As the Hungarian theoretician László Gazdag has argued, these tiny vortices carry information, much as magnetic impulses do on a computer disk. The information carried by a given vortex corresponds to the magnetic momentum of the particle that created it: it is information on the state of that particle. These minute spinning structures travel through the vacuum, and they interact with each other. When two or more of these torsion waves meet, they form an interference pattern that integrates the strands of information on the particles that created them. This interference pattern carries information on the entire ensemble of the particles.

In a simplified but meaningful way we can say that *vacuum vortices record information on the state of the particles that created them—and their interference pattern records information on the ensemble of the particles of which the vortices have interfered.* In this way the vacuum records and carries information on atoms, molecules, macromolecules, cells, even organisms and populations and ecologies of organisms. There is no evident limit to the information that interfering vacuum torsion waves can conserve and convey. In the final count, they can carry information on the state of the whole universe. Throughout the universe, particles are linked by the vacuum in much the same way as objects are linked in the sea: by making and receiving waves.

Consider the interconnections created by the sea. A moment's reflection will tell you that the waves that propagate in the sea produce a real even if temporary connection among the vessels, fish, and other objects that generated them. When a ship travels on the sea's surface, waves spread in its wake. These waves affect the motion of other ships—something that has been dramatically brought home to anyone who has ever sailed a small craft next to an ocean liner. Vessels that are entirely immersed in the sea affect things not only on the surface, but also above and below. A submarine, for example, creates subsurface waves that propagate in every direction. Another submarine—and every fish, whale, or object in the sea—is exposed to these waves and is in a sense shaped, "in-formed," by them. A second submarine likewise "makes waves," and this affects—in-forms—the first, as well as all other things in that part of the sea.

When many things move simultaneously in a waving medium, be it the ordinary sea or the extraordinary vacuum, that medium becomes modulated: full of waves that intersect and interfere. This is what happens when several ships ply the sea's surface. If we view the sea from a height—a coastal hill or an airplane—on a calm day, we can see the traces of ships that passed hours before on that stretch of water. We can also see how the traces intersect and create complex patterns. The modulation of the sea's surface by the ships that disturb it carries information on the ships that created the disturbance. This has practical applications: one can deduce the location, speed, and even the tonnage of the vessels by analyzing the resulting wave-interference patterns.

As fresh waves superimpose on those already present, the sea becomes more and more modulated—it carries more and more information. On calm days its surface remains modulated for hours, and sometimes for days. The wave patterns that persist are the memory of the ships that moved about on that stretch of water. If wind, gravity, and shorelines did not cancel these patterns, this memory would persist indefinitely. But wind, gravity, and shorelines do come into play, and sooner or later the sea's memory dissipates. (This, we should note, does not mean that the memory of the *water* disappears. Water has a remarkable capacity to register and conserve information, as indicated by, among other things, homeopathic remedies that remain effective even when not a single molecule of the original substance remains in a dilution.) In the vacuum, however, there are no forces or things that could cancel or even attenuate waves: the vacuum is considered to be a frictionless medium. In a frictionless medium waves and objects move without resistance and—in the absence of contrary forces—could move forever. Thus, if the vacuum is a truly frictionless medium, the wave memory of the universe may be eternal.

But could any medium be truly frictionless? The answer is yes:

supercooled helium is entirely frictionless, as the Dutch physicist Kammerlingh Onnes discovered in 1911. He took helium—normally a gas—and cooled it degree by degree until it approached the absolute zero of temperature signified by zero on the Kelvin scale. When the temperature of the helium reached 4.2 Kelvin, a dramatic change occurred. Helium lost its gaseous properties: it became liquid. At the same time, under equal pressure, it became 800 times denser! When Onnes cooled this superdense liquid helium still further, at 2.17 Kelvin another major change occurred: the liquid helium became superfluid. Supercooled helium, though it is superdense, does not resist objects passing through it. It flows frictionless through cracks and apertures so tiny that nothing else, not even a much thinner gas, can penetrate them—at least, not without notable friction.

Superfluid helium is a good analogy for the superdense and at the same time frictionless cosmic vacuum. According to John Wheeler's calculations, the energy density of the vacuum is 10^{94} erg per cubic centimeter—a stupendous amount that is far greater than the energy associated with all the matter particles throughout the universe. (Matter particles are particles that have mass and, as Einstein's famous equation tells us, mass accelerated to the square of the velocity of light is equivalent to energy.) The fact is that the vacuum is both superfluid and superdense—much like helium near the absolute zero of temperature. This is a mind-boggling combination, for how can something be denser than anything else and at the same time more fluid than anything else? The vacuum, just like supercooled helium, may be a mind-boggling medium, but it is not a supernatural one.

All things in the universe are immersed in the superdense yet superfluid cosmic vacuum, and all things produce waves that move the vacuum out of its "ground state" (i.e., create vortices that "excite" the vacuum). These torsion waves propagate in the vacuum and they interfere. The interference patterns they create integrate the information carried by the individual vortices. As the vortices of individual things merge, the information they carry is not overwritten, for the waves superpose one on the other. And the superposed waves are in a sense everywhere throughout the vacuum. This, too, is a natural phenomenon: it is familiar in the form of holograms.

In a holographic recording—created by the interference pattern of two light beams—there is no one-to-one correspondence between points on the surface of the object that is recorded and points in the recording itself. Holograms carry information in a distributed form, so all the information that makes up a hologram is present in every part of it. The points that make up the recording of the object's surface are present throughout the interference patterns recorded on the photographic plate: in a way, the image of the object is enfolded throughout the plate. As a result, when any small piece of the plate is illuminated, the full image of the object appears, though it may be fuzzier than the image resulting from illuminating the entire plate.

Superposed vacuum-interference patterns are nature's "holograms"; they carry distributed information on all the particles, and on all the ensembles of particles, throughout the reaches of space and time. The hypothesis we can now advance may be daring, but it is logical. *The quantum vacuum generates the holographic field that is the memory of the universe.*

FIVE

Enter the Akashic Field

All along our review of the puzzles of the mainstream sciences, we have been suspecting that the mysterious field implied by time- and space-transcending correlations in cosmos and consciousness may be an information field at the very heart of the cosmos. This suspicion has been borne out: the zero-point field of the quantum vacuum is not only a superdense energy field; it is also a super-rich information field—the holographic memory of the universe. This finding recalls Indian philosophy's concept of the Akashic Chronicle, the record of everything that happens in the world traced in the Akashic Field. It makes sense to name the newly (re)discovered information field of the universe the "A-field," after ancient tradition's Akashic Field. The A-field takes its place among the fundamental fields of the universe, joining science's G-field (the gravitational field), EM-field (the electromagnetic field), and the various nuclear and quantum fields.

The Akashic Field may be an age-old intuition shared by countless generations, but the field named after it is a radical innovation in contemporary science. We should examine the grounds for this innovation, to make sure that it is not just a chimera of the imagination.

WHY THE A-FIELD—REVIEWING THE EVIDENCE*

The evidence for a cosmic information field—like the evidence for all fundamental laws and processes in nature—is not direct; it must be

---

*Readers more interested in the effects and meaning of the A-field than in the evidence for it can go directly to the next chapter without losing the thread.

reconstructed by reasoning. Like the G-field and the EM-field, the A-field cannot be seen heard, touched, tasted, or smelled. It is indicated, however, by many things that we can and do perceive. These things are not accounted for in the mainstream theories; to the conservative core of the science establishment, they are puzzling and mysterious. Yet the puzzles and mysteries have a common thrust. We can see what this thrust is when we review the bold yet rigorously argued hypotheses—the "scientific fables"—advanced today by cutting-edge investigators in fields as diverse as cosmology, quantum physics, biology, and consciousness research.

Let us revisit, then, the puzzles we encountered in chapter 3 and bring them together with the fables that attempt to throw light on them.

We begin with the puzzles of the universe—the "cosmic puzzles"—and the fables of the Metaverse. We then move to the puzzles at the roots of physical reality—the "quantum puzzles"—and the fables of entanglement and nonlocality that address them. We go next to the puzzles of the living organism and fables about the interconnected web of life. We conclude our review with the puzzles and fables that come to light in the most intimately known domain of our experience: the domain of consciousness.

1. COSMOLOGY

Cosmic Puzzles: Footprints of the A-Field in the Physical Universe

As noted in chapter 3, the standard model of the universe is not as established today as it was even a few years ago. A number of anomalies have come to light, cosmic puzzles that the Big Bang theory cannot explain.

The flat universe. Until the results of observations made with a balloon-based telescope lofted over Antarctica in 1998 became available, cosmologists could not answer the question of whether the universe is flat (with a space-time structure that is essentially "Euclidean"—that is, where light, except near massive bodies, travels in a straight line), or open (with an infinitely expanding negatively curved space-time), or closed (expanding to a limit and then contracting with a positively curved space-time). The correct answer depends on the amount of matter in the universe. If there is more matter in the universe than the "critical density" (estimated at 5 x 10^{-26} g/cm^3), the gravitational pull associated with matter particles will ultimately exceed the inertial force generated by the Big Bang. Then the expansion of the universe will reverse and we will find ourselves in a closed universe that collapses back on itself. If, however, matter density is below the critical quantity, its gravitational pull is more modest, and the force of expansion will continue to dominate it; we then live in an open universe, one that expands forever. If matter density is precisely at the critical value, the forces of expansion and contraction balance each other, making our universe flat: forever balanced at the razor's edge between the opposing forces of expansion and contraction.

Whether the universe is open, closed, or flat seems to have been satisfactorily answered by a number of increasingly sophisticated cosmic probes. First came the Boomerang project's observations of the cosmic microwave background in 1998 ("Boomerang" stands for Balloon Observations of Millimetric Extragalactic Radiation and Geophysics), then the observations of MAXIMA (Millimeter Anisotropy Experiment Imagining Array) and of DASI (Degree Angular Scale Interferometer, based on a microwave telescope at the South Pole). In February of 2003, the findings of the WMAP were released. (The acronym stands for Wilkinson Microwave Anisotropy Probe, which is a satellite launched in Earth orbit in June 30, 2001, recording cosmic radiation from a point on the far side of the moon.) They held no surprises, but refined the previous estimates and provided greater certainty of their validity.

It is now beyond all reasonable doubt that we live in a flat universe. This confirms predictions flowing out of the Big Bang theory, but it is astounding just the same. Because if the universe is flat today, the Big Bang that produced all matter in it must have been fine-tuned to the staggering order of 1 part in 10^{50}. A deviation even of that minute order would have produced an infinitely expanding ("open") or a finite recollapsing ("closed") universe. How this level of precision could have come about has no explanation in the Big Bang theory. That it occurred purely by chance is plausible only if there are a very large number of universes in the cosmos, because then even an improbably well-tuned universe such as ours has some probability of coming about, just as in a very large number of throws of a die even a run of sixes has some likelihood of turning up.

The missing mass. A still more vexing puzzle is why observations through optical telescopes fail to locate the amount of matter we should find in cosmic space. According to current observations, the matter density of the universe is less than 10^{-30} g/cm³—a density that is not sufficient to counteract the force of expansion and create a flat universe. Astrophysicists theorize, however, that a great deal of the matter in the universe is optically invisible. (Visible matter is composed mainly of protons and neutrons, so-called baryons.) Only four percent of the material substance of the universe is made up of objects of visible matter, such as galaxies, stars, planets, interstellar dust, and other astronomical bodies disclosed by optical telescopes. A further twenty-three percent seems to consist of baryonic dark matter (protons and neutrons in structures that are too dim to be visible), as well as of nonbaryonic dark matter (exotic particles such as axions, neutrinos with mass, and WIMPs—weakly interacting massive particles). Yet even the sum total of visible and invisible matter leaves some seventy-three percent of the substance of the universe unaccounted for. This enormous quantity appears to be not matter at all, but "dark energy"—a property of space itself, very likely due to the fluctuation of virtual particles in the quantum vacuum.

Accelerating expansion. In a flat universe—possessing the critical matter density where the inertial force of expansion is precisely balanced by the force of gravitation—the galaxies should be expanding in a gradually slowing fashion, with the momentum of the explosion that drove them apart being progressively slowed by gravitational attraction pulling them toward each other. But this is not the case: the expansion of the galaxies is actually speeding up!

Observations of sufficient precision to determine the expansion of distant galaxies became available only recently. Previously, Edwin Hubble and other astronomers estimated distances to the observed galaxies by assuming that all galaxies have uniform brightness. If so, those that appear brighter would be closer than those that are dimmer. This, however, fails to take into account that there are galaxies with stars of different intrinsic luminosity. It also fails to account for galaxies so far away that the light that reaches us now was emitted in an early phase of their evolution, at which time their intrinsic brightness was considerably different from their brightness as mature galaxies. What astronomers need are galaxies of well-determined brightness, so-called standard candles. By the 1990s, some candles of this kind became known. They are one variety of supernova (the explosion that marks the end of the life cycle of certain stars), known as type Ia.

When a star has reached the stage when it has converted most of the hydrogen in its mass to helium, carbon, oxygen, neon, and some other heavy elements, its outer layers are compressed by gravity to a size roughly the size of Earth but a million times more dense than ordinary matter. Most of these "white dwarfs" cool and fade without dramatic changes, but if one of these superdense objects orbits near an active star, its intense gravity siphons off matter from that star. This increases the white dwarf's density until a thermonuclear chain reaction gets under way. We then have a supernova: the white dwarf explodes, spewing forth its atomic matter at the speed of 10,000 kilometers per second. Since the duration of the supernova depends on its brilliance, astronomers following its evolution can determine its inherent brightness to a high degree of precision.

Dozens of these standard candles have now been studied at distances between four and seven billion light-years away. Their intrinsic brightness can be calculated on the basis of their distance. But these candles are dimmer than their distance would warrant—the observed values do not match the predicted values. This means that they are more distant than the standard model predicts. The cosmos must be expanding faster than cosmologists have thought. Something—some force or energy—must be pushing the galaxies apart.

The current finding brings back the notion of a cosmological constant, first postulated but then discarded by Einstein. In Einstein's "steady state" universe, matter is not created in the womb of a Big Bang but instead is spread homogeneously in space. That it stays so—rather than clumping together under gravitational attraction—is ensured by his cosmological constant, which stands for a force of repulsion that precisely balances the attractive force of gravitation. Consequently, the universe neither expands nor contracts: it remains in a steady state.

Within five years of putting forward the hypothesis of the cosmological constant, Einstein abandoned it, calling it his biggest blunder. Evidence came to light that the universe is unstable, and in a 1923 letter to the mathematician Hermann Weyl, Einstein admitted that if there is no quasi-static world, then one must do away with the cosmological term.

This conclusion was premature. Current measurements of the cosmic background indicate that even if all matter in the universe originated in a Big Bang, space-time is nevertheless flat: the universe should be precisely balanced between expansion and contraction. Yet the galaxies are expanding! Perhaps there is a cosmological constant after all, one that pushes apart the cosmos, rather than just keeping it in a steady state.

Cosmologists suspect that the quantum vacuum is the source of the strange energies represented by this constant. Space is filled with virtual particles, in constant fluctuation. The energy of the particles matches the effects attributed to them, even if they themselves exist too briefly to be measured. This energy—the positive cosmological constant—is believed to be responsible for the accelerating expansion of the galaxies. The assumption is not new: already in the 1960s, the physicist Yakov Zeldovich showed that vacuum energies act in precisely the way presupposed in Einstein's estimate of the cosmological constant.

But this assumption is not perfect: the sum total of the energy content of the quantum vacuum is far greater than the value required for the additional force of expansion. As John Wheeler's calculations show, the magnitude of vacuum energy is mind-boggling—even when corrections due to quantum effects are taken into account, it still exceeds by some 120 orders of magnitude the energy contained in all matter throughout the universe! (The vacuum energy that sets the cosmological constant should be less than 10^{-8} joules per cubic meter, but the most reasonable calculation of vacuum energy yields a value of 10^{112} joules per cubic meter—which is *10^{120} times too much.*) Because gravitation is associated with energy (as defined in Einstein's formula $E = mc^2$), this excess energy would inject so much gravitation into the universe that particles would accelerate even in the absence of other objects, and all things made of particles (planets, stars, galaxies) would fly apart. The universe would expand like a rapidly inflating balloon. In every region of space, the cosmological constant would dramatically thin the matter content of the cosmos. In our vicinity, space would be nearly empty. Looking up at the night sky, we would not see anything other than the moon and the planets of our solar system. Indeed, we would not see even those: assuming that general relativity theory holds true, space-time would be so highly curved that visibility would be limited to a single kilometer. In the daytime we would not see the Sun, not even planes flying higher than a thousand meters. Yet we see the Sun and high-flying planes in the daytime, and billions of stars billions of light-years away at night.

Obviously there is something in the universe, some factor or combination of factors, that keeps the cosmological constant, if not at zero, at the small but precise positive excess value that makes for the observed expansion of the galaxies without blowing the universe apart.

Coherence of cosmic ratios. There are a number of strange coincidences regarding the observed parameters of the universe. Already in the 1930s, Sir Arthur Eddington and Paul Dirac noted some remarkable facts about the "dimensionless ratios" that relate the universe's basic parameters to each other. For example, the ratio of the electric force to the gravitational force is approximately 10^{40}, and the ratio of the observable size of the universe to the size of elementary particles is likewise around 10^{40}. This is all the more strange as the former ratio should be unchanging (the two forces are assumed to be constant), whereas the latter is changing (since the universe is expanding). In his "large number hypothesis," Dirac speculated that the agreement of these ratios, the one variable, the other not, is not merely a temporary coincidence. But if the coincidence is more than temporary, either the universe is not expanding or the force of gravitation varies in accordance with its expansion!

Additional coincidences involve the ratio of elementary particles to the Planck-length (this ratio is 10^{20}) and the number of nucleons in the universe ("Eddington's number," which is approximately 2×10^{79}). These are very large numbers, yet "harmonic" numbers can be constructed from them. For example, Eddington's number is roughly equal to the square of 10^{40}.

Recently the astrophysicist Menas Kafatos together with Robert Nadeau and Roy Amoroso showed that many of these coincidences can be interpreted in terms of the relationship on the one hand between the masses of elementary particles and the total number of nucleons in the universe, and on the other between the gravitational constant, the charge of the electron, Planck's constant, and the speed of light. Scale-invariant relationships appear—the physical parameters of the universe turn out to be proportional to its overall scale.

The "horizon problem." The coherence indicated by numerical relationships is reinforced by observational evidence. The latter gives rise to the so-called horizon problem: the problem of the large-scale uniformity of the cosmos at all points of the horizon as seen from Earth. This comes to the fore both in regard to the universe's background radiation and in relation to the evolution of its galaxies.

The universe's microwave background radiation proves to be isotropic (the same in all directions). This radiation is believed to be the remnant of the Big Bang; according to BB theory, it was emitted when the universe was about 400,000 years old. The problem is that at that point in time the opposite sides of the expanding universe were already ten million light-years apart. By that time, light could have traveled only 400,000 light-years, so no physical force or signal could have connected the expanding regions. Yet the cosmic background radiation is uniform for billions of light-years wherever we look in space.

This is true not only of the background radiation; galaxies and multi-galactic structures in the cosmic "foreground" also evolve in a uniform manner in all directions from Earth. This is the case even in regard to galaxies that have not been in physical contact with each other since the birth of the universe. If a galaxy ten billion light-years from Earth in one direction exhibits structures analogous to a galaxy the same distance away in the opposite direction, then structures that are twenty billion light-years from each other are uniform. This uniformity cannot be the consequence of physical linkages, since the highest rate at which physical forces can propagate in space-time is the speed of light. Although by now light reached across the ten-billion-light-year distance to Earth from each of the galaxies (which is why we can see them), in a universe less than twenty billion years old it could not have reached from one of these galaxies to the other. Nonetheless, even over distances not connected by light, our 13.7-billion-year-old universe evolves as a coherent whole.

The tuning of the constants. Perhaps the most mysterious of all the cosmic puzzles is the observed "fine-tuning" of the physical constants of the universe. The basic parameters of the cosmos have precisely the value that allows complex structures to arise. From our perspective this is fortunate, for the existence of these structures is a precondition of life on this planet—if the universe were any less finely tuned, we would not be here to speculate on the reasons for this precision. But is this mere serendipity?

The fine-tuning in question involves upward of thirty factors and considerable accuracy. For example, if the expansion rate of the early universe had been one-billionth less than it was, the universe would have re-collapsed almost immediately; and if it had been one-billionth more, it would have flown apart so fast that it could produce only dilute, cold gases. A similarly small difference in the strength of the electromagnetic field relative to the gravitational field would have prevented the existence of hot and stable stars like the Sun, and hence the evolution of life on planets associated with these stars. Moreover, if the difference between the mass of the neutron and the proton were not precisely twice the mass of the electron, no substantial chemical reactions could take place, and if the electric charge of electrons and protons did not balance precisely, all configurations of matter would be unstable and the universe would consist of nothing more than radiation and a relatively uniform mixture of gases.

But even the astonishingly precisely adjusted laws and constants do not fully explain how the universe would have evolved out of the primordial radiation field. Galaxies had formed out of this radiation field when the expanding universe's temperature dropped to 3,000 degrees on the Kelvin scale. At that point the existing protons and electrons formed atoms of hydrogen, and these atoms condensed under gravitational pull, producing stellar structures and the giant swirls that make for the birth of galaxies. Calculations indicate that a very large number of atoms would have had to come together to start the formation of galaxies, perhaps of the order of 10^{16} suns. It is by no means clear how this enormous quantity of atoms—equivalent to the mass of 100,000 galaxies—would have come together. Random fluctuations among individual atoms do not furnish a plausible explanation.

Cosmic Fables: The Universe of Universes

The rapidly growing field of physical cosmology is full of puzzles— anomalies that the established theories cannot explain. But cosmologists are not stumped. In recent years a number of "cosmic fables" have seen the light of day, including those that argue that *our* universe is not all there is in the world. The larger reality, these new "cosmological scenarios" tell us, is the Metaverse, the mother of our universe and perhaps of a vast number of other universes. Metaverse scenarios, as we have noted in chapter 3, deserve serious attention: they offer a particularly promising approach to the puzzles that beset contemporary cosmology.

SOME CURRENT
METAVERSE SCENARIOS

A widely discussed scenario advanced by the Princeton physicist John Wheeler claims that the expansion of the universe will come to an end, and ultimately the universe will collapse back on itself. Following this "Big Crunch," it could explode again, giving rise to another universe. In the quantum uncertainties that dominate the supercrunched state, almost infinite possibilities exist for universe creation. This could account for the fine-tuned features of our universe since, given a sufficiently large number of successive universe-creating oscillations, even the improbable fine-tuning of a universe such as ours has a chance of coming about.

It is also possible that many universes come into being at the same time. This, in turn, is the case if the explosion that gave rise to them was "reticular"—made up of a number of individual regions. In the Russian-born cosmologist Andrei Linde's inflation theory, the Big Bang had distinct regions, much like a soap bubble in which smaller bubbles are stuck together. As such a bubble is blown up, the smaller bubbles become separated and each forms a distinct bubble of its own. The bubble universes percolate outward and follow their own evolutionary destiny. Each bubble universe hits on its own set of physical constants, and these may be very different from those of other universes. For example, in some universes gravity may be so strong that they recollapse almost instantly; in others, gravity may be so weak that no stars could form. We happen to live in a bubble tuned in such a way that complex systems, including humans, could evolve in it.

New universes could also be created inside black holes. The extreme high densities of these space-time regions represent "singularities" where the known laws of physics do not apply. Stephen Hawking and Alan Guth suggested that under these conditions the black hole's region of space-time detaches itself from the rest and expands to create a universe of its own.

In another scenario, baby universes are periodically created in bursts similar to that which brought forth our own universe. The QSSC (Quasi-Steady State Cosmology) advanced by Fred Hoyle together with George Burbidge and J. V. Narlikar postulates that such "matter-creating events" are interspersed throughout the meta-universe. Matter-creating events come about in the strong gravitational fields associated with dense aggregates of preexisting matter, such as in the nuclei of galaxies. The most recent burst occurred some fourteen billion years ago, in excellent agreement with the latest observations regarding the age of our own universe.

Yet another Metaverse scenario is the work of Ilya Prigogine and his colleagues J. Geheniau, E. Gunzig, and P. Nardone. Their theory agrees with the QSSC in suggesting that major matter-creating bursts similar to our Big Bang occur from time to time. The large-scale geometry of space-time creates a reservoir of "negative energy" (which is the energy required to lift a body away from the direction of its gravitational pull); from this reservoir, gravitating matter extracts positive energy. Thus gravitation is at the root of the ongoing synthesis of matter: it produces a perpetual matter-creating mill. The more particles are generated, the more negative energy is produced and then transferred as positive energy to the synthesis of still more particles. Given that the quantum vacuum is unstable in the presence of gravitational interaction, matter and vacuum form a self-generating feedback loop. A critical matter triggered instability causes the vacuum to transit to the inflationary mode, and that mode marks the beginning of a new era of matter synthesis.

A recent Metaverse scenario is the work of Paul J. Steinhardt, of Princeton, and Neil Turok, of Cambridge. Their cosmology accounts for all the facts accounted for by the Big Bang theory and also gives an explanation of the puzzling accelerating expansion of distant galaxies. According to Steinhardt and Turok, the universe—which is effectively the Metaverse—undergoes an endless sequence of cosmic epochs, each of which begins with a "Bang" and ends in a "Crunch." Each cycle includes a period of gradual and then further accelerating expansion, followed by reversal and the beginning of an epoch of contraction. They estimate that at present we are about fourteen billion years into the current cycle and at the beginning of a trillion-year period of accelerated expansion. Ultimately our universe (which is our cycle of the Metaverse) will achieve the condition of homogeneity, flatness, and energy needed to begin the next cycle. In this model the Metaverse is infinite and flat, rather than finite and closed, as in the oscillating universe models.

The great variety of cosmological scenarios advanced today indicates on the one hand that there is no definitive consensus yet regarding the birth and evolution of our universe. But on the other hand it tells us that fables of the Metaverse make sense: it is entirely reasonable to believe that this universe is not "all there is." There is also a meta-universe that is the originating ground, the quasi-permanent and possibly infinite womb of the universe we observe and inhabit.

Metaverse cosmologies have enormous explanatory potential. They can explain in principle how our universe came by the remarkable properties it actually has. Such an explanation is needed, because a universe such as ours—with galaxies and stars, and life on this and presumably other life-supporting planets—is not likely to have come about as a matter of serendipity. According to Roger Penrose's calculations, the probability of hitting on our universe by a random selection from among the alternative-universe possibilities is one in $10^{10^{123}}$. This is an inconceivably large number, indicating an improbability of astronomical dimensions. Indeed, Penrose himself speaks of the birth of our universe as a "singularity" where the laws of physics do not hold.

But if our particular universe is so staggeringly improbable, how did it come about? The explanation we can derive from Metaverse cosmologies is simple and powerful. We know that the vacuum fluctuations that preceded the birth of our universe were precisely such that a life-bearing universe could come about. We also know that these fluctuations were not created by the primal explosion known as the Big Bang—that stupendous event only *amplified* them. The fluctuations that led to our staggeringly coherent universe were already present when our universe was born; they were there in its vacuum "pre-space." In light of the new Metaverse cosmologies, we need not assume that they were there as a matter of pure serendipity, nor do we have to apply to a transcendental force or agency for selecting them. As we shall discuss in the next chapters, the selection of just the right fluctuations was very likely due to the information conveyed to our universe from a prior universe. This is perfectly plausible, given that the cosmic vacuum was the womb of our universe, and that it was modulated by universes that preceded ours. The A-field, it appears, not only creates coherence in our universe, but also links our universe with prior universes in the Metaverse.

2. QUANTUM PHYSICS

Quantum Puzzles: Traces of the A-Field at the Roots of Reality

At the beginning of the twentieth century, new observations and experiments raised questions about the most fundamental assumptions of Newton's classical mechanics. Although the laws of motion advanced by Newton continue to hold true under conditions at the surface of Earth, the fundamental nature of the universe cannot be accommodated under the heading of the classical conceptions. Space proves to be more than a passive receptacle, and time does not flow equitably through all eternity. Space and time have been joined by Einstein into a four-dimensional continuum, and this continuum interacts with the events—the particles of matter and light—that move about in it.

Einstein's relativity revolution took place in the first decade of the twentieth century, and some twenty years later another revolution occurred: the quantum revolution. This was just as fundamental as that triggered by Einstein. Relativity theory did away with space and time as the backdrop for the deterministic motion of mass points, but it preserved the unambiguous description of the basic entities of the physical universe. Quantum theory, on the other hand, did away with unambiguous paths of motion (particles no longer appeared to move in just one determinate way but seemed to move in a way that allows a choice between alternative motions), and introduced indeterminacy into the very foundations of reality (a level of freedom—or randomness—in determining just which path a particle would follow). The mechanistic, predictable world of classical mechanics became fuzzy. It was replaced by a strange world that Heisenberg, Bohr, and other quantum physicists refused to interpret in realistic terms.

Superposed wave state. The quanta of light and energy that emerged from ever more sophisticated experiments refused to behave as tiny equivalents of familiar objects. Their behavior proved more and more weird. Though Einstein received the Nobel Prize for his work on the photoelectric effect (where streams of light quanta are generated on irradiated plates), he did not suspect—and was never ready to accept—the strangeness of the quantum world. But physicists investigating the behavior of these packets of light and energy found that, until an instrument of detection or another act of observation registers them, they do not have a specific position, nor do they occupy a unique state. The ultimate units of physical reality have no uniquely determinable location, and they exist in a strange state that consists of the simultaneous "superposition" of several ordinary states.

Newton's mass points and Democritus's atoms could be unambiguously defined in terms of force, position, and motion, but the quantum cannot. Its description is complex and intrinsically ambiguous. It exists in several states at the same time; this is expressed by the particle's "wave function"—the mathematical description that relates its superposed wave state to its classical space-time state. A quantum of light or energy occupies all its states at the same time—in potential. Until it is observed or registered by an instrument, it is indeterminate as to the choice among them. But as soon as it is observed or measured, its weird ability to be in several states at the same time resolves into the normal condition in which a particle is in just one state at any one time. Then, physicists say, its superposed wave function "collapses." When it does, a particle can be described in the classical manner, as an object in a single, determinate state.

Complementarity and uncertainty. Until very recently (for evidence contrary to this tenet has now surfaced), particles were believed to exhibit the property Nils Bohr called "complementarity." Depending on how they were observed and measured, particles were said to be either corpuscles or waves, but not both at the same time. The alternative properties of particles were held to be complementary: although they do not appear singly, together they fully describe the state of the particles.

To compound the mystery, the various states of particles cannot all be measured at the same time. If one measures a particle's position, for example, its momentum (which is the product of its mass and velocity) becomes indistinct; and if one measures its momentum, its position becomes blurred. This is known as Heisenberg's principle of uncertainty.

Indeterminacy and randomness. The strangeness of the particle is exacerbated by the way in which its potential states resolve into an actual state. As we have seen, in the pristine state the quantum is in a superposed state where it has neither one distinct location nor a full set of measurable properties. But when it is observed or measured, the quantum's wave function "collapses": its superposed state changes into the classical state, with unique location and full measurability. However, there are no laws of physics that can predict just which of its possible states the particle will occupy. While in the aggregate the collapse of the superposed into the singular state conforms to statistical rules of probability, there is no way to tell just how it will unfold in a given instance. Unless each outcome of each wave-function collapse takes place in a separate universe (as Everett suggested), individual multiple-state resolutions are indeterministic "quantum jumps" that are not subject to any law of physics.

Einstein was opposed to the fundamental role of chance in nature—he said: "God doesn't play dice." Something is missing in the observational and theoretical arsenal of quantum mechanics, he suggested; in some essential respects the theory is incomplete. But Bohr countered that the very question of what a particle is "in-itself" is not meaningful and should not even be asked. Eugene Wigner echoed this view when he said that quantum physics deals with *observations*, and not with *observables*. Heisenberg also supported it when he spoke of the error of the "philosophical doctrine of Democritus," which claims that the whole world is made up of objectively existing material building blocks called atoms. The world, said Heisenberg, is built as a mathematical, and not as a material, structure. In consequence there is no use asking to what the equations of mathematical physics refer—they do not refer to anything beyond themselves.

Quantum Fables: Entanglement and Nonlocality

The physicist David Bohm was among the first to refuse to accept the weird behavior of the quantum as a full description of reality. His "hidden variables theory" suggested that the selection of the state of the quantum is not random, but rather guided by real physical processes. He theorized that a pilot wave called the quantum potential "Q" emerges from a deeper, unobservable domain of the universe and guides the observed behavior of particles. Thus, particle behavior is weird and indeterministic only at the surface; at the deeper level it is determined by the quantum potential. Later Bohm identified the deeper level of reality as the "implicate order"—a holofield where all the states of the quantum are permanently coded. Observed reality emerges from this field by constant unfolding: it is the "explicate order."

Various versions of Bohm's theory are being developed today by theoretical physicists who are unwilling to take the mathematical formalisms of quantum physics for an adequate explanation of the real world. They account for the behavior of the quantum in reference to its interaction with a deeper dimension of the multidimensional space-filling field that has now replaced the "luminiferous ether" of the nineteenth century.

This is a relatively recent development. Until the 1980s, quantum weirdness was generally accepted as an irreducible condition of the ultrasmall domain of the universe. Physicists contented themselves with the smooth functioning of the equations by which they computed their observations and made predictions. In the last two decades the picture has begun to change. With the new fables, a far less weird view of the quantum world is beginning to take shape. Experiments that were originally designed to investigate the complementary corpuscular/wave nature of the quantum have been instrumental in bringing about the new understanding.

The first experiment to demonstrate the wave nature of light was conducted by Thomas Young in 1801. In his famous "double-slit experiments," coherent light was allowed to pass through a filtering screen with two slits. (Young created coherent light by making a ray of sunlight penetrate a pinhole; today, lasers are used for this purpose.) When Young placed a second screen behind the filter with the two slits, he found that instead of two pinpoints of light, a wave-interference pattern appeared on the screen. The same effect can be observed on the bottom of a pool when two drops or pebbles disturb the sunny and otherwise smooth surface of the water. The waves spreading from each disturbance meet and interfere with each other: where the crest of one wave meets the crest of the other, they reinforce each other and appear bright. Where crest meets trough they cancel each other and appear dark.

Are the quanta that pass through Young's slits waves? If so, they could then pass through both slits and form interference patterns. This assumption makes sense until such a weak light source is used in the experiments that only one photon is emitted at a time. Commonsense reasoning tells us that a single photon cannot be a wave: it must be a corpuscular packet of energy of some sort. But then it should be able to pass through only one of the slits and not both slits at the same time. Yet when single photons are emitted, a wave-interference pattern builds up on the screen, as if each photon passed through both slits.

The "split-beam" experiment, designed by John Wheeler, discloses the same dual effect. Here, too, photons are emitted one at a time, and they are made to travel from the emitting gun to a detector that clicks when a photon strikes it. A half-silvered mirror is inserted along the photon's path, which splits the beam. This means that on the average, one in every two photons will pass through the mirror and one in every two will be deflected by it. To verify this, photon counters are installed both behind the half-silvered mirror and at right angles to it. There is no problem here: the two counters register an approximately equal number of photons. But a curious thing occurs when a second half-silvered mirror is inserted in the path of the photons that are undeflected by the first. One would still expect that an equal number of photons would reach the two counters: deflection by the two mirrors would simply have exchanged their individual destinations. But this is not the case. One of the two counters registers all the photons—none arrives at the other.

It appears that the kind of interference that was noted in the double-slit experiment occurs in the split-beam experiment as well, indicating that individual photons are behaving as waves. Above one of the mirrors the interference is destructive (the phase difference between the photons is 180 degrees), so that the wave patterns of the photons cancel each other. Below the other mirror the interference is constructive (since the wave phase of the photons is the same) and in consequence the photon waves reinforce each other.

The interference of wave patterns of photons emitted moments apart in the laboratory has also been observed in photons emitted at considerable distances from us, at considerable intervals of time. The "cosmological" version of the split-beam experiment bears witness to this. In this experiment the photons are emitted not by an artificial light source, but by a distant star. In one case the photons of the light beam emitted by the double quasar known as 0957+516A,B were tested. This distant "quasi-stellar object" appears to be two, but is in fact one and the same object, its double image being due to the deflection of its light by an intervening galaxy situated about one fourth of the distance from Earth. (The presence of mass, according to relativity theory, curves space and hence also the path of the light beams that propagate in it.) A light beam taking the curved path takes longer to travel than one coming by the straight path. In this case the additional distance traveled by the light deflected by the intervening galaxy means that the photons that make up the deflected beam have been on the way fifty thousand years longer than those that traveled by the more direct route. Although originating billions of years ago and arriving with an interval of fifty thousand years, the photons of the two light beams interfere with each other just as if they had been emitted seconds apart in the laboratory.

Repeatable and indeed oft repeated experiments show that—whether they are emitted at intervals of a few seconds in the laboratory or at intervals of thousands of years elsewhere in the universe—particles that originate from the same source interfere with each other. Is a photon or an electron a corpuscle when emitted (since it can be emitted one by one) and a wave when it propagates (since it produces wavelike interference patterns when it encounters other photons or electrons)? And why does the coupling of this particle wave persist almost infinitely, even over cosmological distances? The search for an answer to these questions points in a new direction.

Recent versions of the double-slit experiment furnish an indication of the direction in which the answer is now being sought. Initially the experiments were designed to answer a simple question: Does the particle really pass through both slits, or only one? And if only one, which one? The experiment consists of an apparatus that allows each photon access to only one of the two slits. When a stream of photons is emitted and confronted with the two slits, the experiment should decide which of the slits a given photon is passing through.

In accordance with Bohr's principle of complementarity, when the experiment is set up so that the path of the photons can be observed, the corpuscular face of the photons appears and the wave-face disappears:

the interference fringes diminish and can entirely vanish. The higher the power of the "which-path detector," the more the interference fringes diminish. This was shown by an experiment conducted by Mordehai Heiblum, Eyal Buks, and colleagues at Israel's Weizmann Institute. Their state-of-the-art technology comprised a device less than one micrometer in size, which creates a stream of electrons across a barrier on one of two paths. The paths focus the electron streams and enable the investigators to measure the level of interference between the streams. The higher the detector is tuned for sensitivity, the less there is of interference. With the detector turned on for both paths, the interference fringes disappear.

This result conforms to Bohr's theory, according to which the two complementary faces of particles can never be observed at one and the same time. However, an ingenious experiment by Shahriar Afshar, a young Iranian-American physicist, demonstrated that even when the corpuscular face is observed, the wave-aspect is still there: *the interference pattern does not disappear.* In this experiment, reported in July 2004 by the British journal *New Scientist*, a series of wires are placed precisely where the dark fringes of the interference pattern should be. When light hits the wires, they scatter it so that less light reaches the photon detector. But light does not affect these particular points: even when photons pass through the slits one at a time, the dark fringes are still in place.

The continued presence of the interference pattern suggests that particles continue to behave as waves even when they are individually emitted; only their wave-face is not apparent to conventional observation. Asfhar suggests—and a number of particle physicists are inclined to agree—that the wave-aspect of the particle is the fundamental aspect. The corpuscular face is not the real face: the entire experiment can be described in terms of photon-waves.

Does this mean that the mysteries surrounding the behavior of particles are resolved? Not by any means. Even as a wave-state, the state of the particle is decidedly non-commonsensical: it is "nonlocal." The "which-path detecting apparatus" appears to be coupled in an instant and non-energetic manner with the photons passing through the slits.

The effect is astonishing. In some experiments the interference fringes disappear as soon as the detector apparatus is readied—and even when the apparatus is not turned on! Leonard Mandel's optical-interference experiment of 1991 bears this out. In the Mandel experiment two beams of laser light are generated and allowed to interfere. When a detector is present that enables the path of the light to be determined, the interference fringes disappear as Bohr predicted. But the fringes disappear *regardless of whether or not the determination is actually carried out.* The very possibility of "which-path-detection" destroys the interference pattern.

This finding was confirmed in the fall of 1998, when University of Konstanz physicists Dürr, Nunn, and Rempe reported on an experiment where interference fringes are produced by the diffraction of a beam of cold atoms by standing waves of light. When no attempt is made to detect which path the atoms are taking, the interferometer displays fringes of high contrast. However, when information is encoded within the atoms as to the path they take, the fringes vanish. The labeling of the paths does not need to be read out to produce the disappearance of the interference pattern; it is enough that the atoms are labeled so that this information *can* be read out.

Is there an explanation for this strange finding? There is, but it is not a commonsense one. It appears that whenever one encodes "directional information" in a beam of atoms, this information correlates the atom's momentum with its internal electronic state. Consequently when an electronic label is attached to each of the paths the atoms can take, the wavefunction of one path becomes orthogonal—at right angles—to the other. And streams of atoms or photons that are orthogonal cannot interfere with each other.

The fact is that atoms, the same as particles, can be nonlocally correlated with each other, and even with the apparatus through which they are measured. In itself, this is not new: nonlocality in the quantum world has been known for more than half a century. Already in 1935 Erwin Schrödinger suggested that particles do not have individually defined quantum states but occupy collective states. The collective superposition of quantum states applies to two or more properties of a single particle, as well as to a set of particles. In each case it is not the property of a single particle that carries information, but the state of the ensemble in which the particle is embedded. The particles themselves are intrinsically "entangled" with each other, so that the superposed wavefunction of the entire quantum system describes the state of each particle within it.

NONLOCALITY:
THE REVOLUTIONARY EXPERIMENTS

THE EPR EXPERIMENT

The EPR experiment—the first of the revolutionary experiments that prove the nonlocality of the microsphere of physical reality— was put forward by Albert Einstein with his colleagues Boris Podolski and Nathan Rosen in 1935. This "thought experiment" (so called because at the time it could not be empirically tested) requires that we take two particles in a so-called singlet state, where their spins cancel out each other to yield a total spin of zero. We then allow the particles to separate and travel a finite distance. If we could then measure the spin states of both particles, we would know both states at the same time. Einstein believed that this would show that the strange limitation specified in Heisenberg's principle of uncertainty is not a complete description of physical reality.

When experimental apparatus sophisticated enough to test this possibility could be devised, it turned out that this is not exactly what happens. Suppose that we measure the spin state of one of the particles—particle A—along some direction, let us say the z-axis (the permissible spin states are "up" or "down" along axes x, y, and z). Let us say we find that this measurement shows the spin to be in the direction "up." Because the spins of the particles have to cancel each other, the spin of particle B must definitely be "down." But the particles are removed from each other; this requirement should not hold. Yet it does. Every measurement on one particle yields a complementary outcome in the measurement on the other. It appears that the measurement of particle A has an instantaneous effect on B, causing its spin wave function to collapse into the complementary state. The measurement on A does not merely reveal an already established state of B: it actually *produces* that state.

An instantaneous effect propagates from A to B, conveying precise information on what is being measured. B "knows" when A is measured, for what parameter, and with what result, for it assumes its own state accordingly. *A nonlocal connection links A and B, notwithstanding the distance that separates them.* Empirical experiments performed in the 1980s by Alain Aspect and collaborators and repeated by Nicolas Gisin in 1997 show that the speed with which the effect is transmitted is mind-boggling: in Aspect's experiments, the communication between particles twelve meters apart was estimated at less than one billionth of a second, about twenty times faster than the speed with which light travels in empty space, while in Gisin's experiment, particles ten kilometers apart appeared to be in communication 20,000 times faster than the velocity of light, relativity theory's supposedly unbreakable speed barrier. The experiments also show that the connection between the particles is not transmitted by conventional means through the measuring apparatus; it is intrinsic to the particles themselves. The particles are "entanged": their correlation is not sensitive either to distance in space or to difference in time.

Subsequent experiments have involved more particles over ever-larger distances (at the time of writing, up to forty-one kilometers), without modifying these surprising results. It appears that separation does not divide particles from each other—otherwise a measurement on one would not produce an effect on the other. It is not even necessary that the particles have originated in the same quantum state, so that they originally formed one system. Experiments show that any two particles, be they electrons, neutrons, or photons, can originate at different points in space and in time; if they once come together within the same system of coordinates, that is enough for them to continue to act as part of the same quantum system even when they are separated . . .

THE TELEPORTATION EXPERIMENTS

Recent experiments show that a form of nonlocal connection known as "teleportation" exists not only between individual quanta, but also between entire atoms. Teleportation has been experimentally proven since 1997 in regard to the quantum state of photons in light beams and the state of magnetic fields produced by clouds of atoms. In the spring of 2004 milestone experiments by two teams of physicists, one at the National Institute of Standards in Colorado and the other at the University of Innsbruck in Austria, demonstrated that the quantum state of entire atoms can be teleported by transporting the quantum bits ("qubits") that define the atoms. In the Colorado experiment led by M.D. Barrett the ground state of beryllium ions was successfully teleported, and in the Innsbruck experiment headed by M. Riebe the ground and metastable states of magnetically trapped calcium ions were teleported. The physicists achieved teleportation of a remarkably high fidelity (78% by the Colorado team and 75% by the Innsbruck team) using different techniques, but following the same basic protocol.

First two charged atoms (ions), labeled A and B, are "entangled," creating the instant link that is also observed in the EPR experiment. Then a third atom, labeled P, is prepared by encoding in it the coherently superposed quantum state that is to be teleported. Then A, one of the entangled ions, is measured together with the prepared atom P. At that point the internal quantum state of B transforms: it assumes the exact state that was encoded in P! It appears that the quantum state of P has been "teleported" to B.

Although the experiments involved complex procedures, the real-world process they demonstrate is basic and straightforward. When A and P are measured together, the preexisting nonlocal connection between A and B creates a nonlocal transfer of state from P to B. In the EPR experiment, one of a pair of entangled particles "in-forms" the other of its measured state; similarly, in teleportation experiments, the measurement of one of a pair of entangled ions together with a third ion encodes the state of the latter in the other twin. Because the process destroys the super-posed quantum state of A and re-creates it in P, it recalls science fiction's idea of "beaming" an object from one place to another.

While beaming entire objects, not to mention people, is far beyond the current realm of possibilities, the equivalent process on the human level can be envisaged. In this "thought experi-ment" we take two persons who are emotionally close to each other, let us say Archie and Betty, young people deeply in love. We ask a third person, Petra, to concentrate on a given thought or image. We then create a deep "transpersonal" connection between Archie and Petra by having them pray or meditate together. If human-level teleportation works, at the very instant Archie and Petra enter a meditative state, the thought or image Petra has been concentrating on vanishes from her mind, and it reappears in the mind of Betty.

State-of-the-art teleportation experiments open vast vistas. Even though we will not realistically be able to beam macroscale objects or people in the foreseeable future, we could learn to beam thoughts and images, and physicists should be able to find ways to beam qubits not just from one atom to another, but among a large number of particles simultaneously. This would be the basis for a new generation of superfast quantum computers. When a large number of entangled particles are distributed through the structure of a computer, it is expected that "quantum teleportation" will create an instant transfer of information among them without requiring that they be wired together or even be next to each other. The quantum computer could also be governed from a distance, although the remote software would have to be disposable—the instant the information it contained appeared in the computer, it would vanish at the remote location.

In the words of physicist Nick Herbert, "[T]he essence of nonlocality is unmediated action-at-a-distance. . . . A nonlocal interaction links up one location with another without crossing space, without decay, and without delay." This linking, according to the quantum theoretician Henry Stapp, could be the "most profound discovery in all of science."

On first sight "action-at-a-distance" is strange (Einstein called it "spooky"), but it is not stranger than many other aspects of the quantum domain. And it is a puzzle only if we fail to recognize the bona fide physical factor that is responsible for it. Recognizing the real-world factor that underlies nonlocality calls for a new paradigm in the sciences, because the interaction involved in nonlocality is not any known form of interaction: it does not involve the expenditure of energy, and it transcends the hitherto known bounds of space and time. Nonlocal interaction is instant "informational" interaction and, as we shall discuss, it is best viewed as the action of a physically real information-field: the A-field.

3. BIOLOGY

Puzzles of the Living State: The A-Field in the Organism

The physical world has turned out to be strange almost beyond belief, but the world of life seems to conserve a measure of commonsense rationality. This is not entirely so, however. The living organism is not a mere biochemical machine. As the experimental biophysicist Mae-Wan Ho pointed out, it is dynamic and fluid, its myriad activities self-motivated, self-organizing, and spontaneous. Local freedom and global cohesion are maximized; part and whole are mutually implicated and mutually entangled.

Whole-system coherence. The coherence of the organism is quintessentially pluralistic and diverse at every level, from the tens of thousands of genes and hundreds of thousands of proteins and other macromolecules that make up a cell, to the many kinds of cells that constitute tissues and organs. There are no controlling and controlled parts or levels; all components are in instant and continuous communication. As a result the adjustments, responses, and changes required for the maintenance of the organism propagate in all directions at the same time. This kind of instant, system-wide correlation cannot be produced solely by physical or even chemical interactions among molecules, genes, cells, and organs. Though some biochemical signaling—for example, of control genes—is remarkably efficient, the speed with which activating processes spread in the body, as well as the complexity of these processes, makes reliance on biochemistry alone insufficient. The conduction of signals through the nervous system, for example, cannot proceed faster than about twenty meters per second, and it cannot carry a large number of diverse signals at the same time. Yet there is evidence that the entire organism is subtly but effectively interlinked; there are quasi-instant, nonlinear, heterogeneous, and multidimensional correlations among all its parts.

No matter how diverse the cells, organs, and organ systems of the organism, in essential respects they act as one. According to Mae-Wan Ho they behave like a good jazz band, where every player responds immediately and spontaneously to however the others are improvising.

The super jazz band of an organism never ceases to play in a lifetime, expressing the harmonies and melodies of the individual organism with a recurring rhythm and beat but with endless variations. Always there is something new, something made up, as it goes along. It can change key, change tempo, or change tune, as the situation demands, spontaneously and without hesitation. There is structure, but the real art is in the endless improvisations, where each and every player, however small, enjoys maximum freedom of expression, while remaining perfectly in step with the whole.

The "music" of a higher organism ranges over more than seventy octaves. It is made up of the vibration of localized chemical bonds, the turning of molecular wheels, the beating of micro-cilia, the propagation of fluxes of electrons and protons, and the flowing of metabolites and ionic currents within and among cells through ten orders of spatial magnitude.

The level of coherence discovered in the organism suggests that in some respects it is a macroscopic quantum system. Living tissue is a "Bose-Einstein condensate": a form of matter in which quantum-type processes, hitherto believed to be limited to the microscopic domain, occur at macroscopic scales. That they do has been verified in 1995, in experiments for which the physicists Eric A. Cornell, Wolfgang Ketterle, and Carl E. Wieman received the 2001 Nobel Prize. The experiments show that under certain conditions, seemingly separate particles and atoms interpenetrate as waves. For example, rubidium and sodium atoms behave not as classical particles but as nonlocal quantum waves, penetrating throughout the given condensate and forming interference patterns.

The system-wide coherence of the organism provides further evidence for the quantum postulate. It is known that correlation can occur between distant molecules and molecular assemblies only when they resonate at the same or compatible frequencies. Whether the force that appears among such assemblies is attractive or repulsive depends on the given phase relations. For cohesion to occur among the assemblies, they have to resonate in phase—the same wave function must apply to them.

This provision applies also to the coupling of frequencies among the assemblies. If faster and slower reactions are to accommodate themselves within a coherent overall process, the respective wave functions must coincide. They do in fact coincide, and as a consequence quantum biologists can speak of a "macroscopic wave function" that applies to the organism as a whole.

Superconductivity. In the living organism, processes suggestive of superconductivity appear at macroscopic scales and normal temperatures. The detailed mechanism underlying these phenomena is the subject of intense research. Hans-Peter Dürr, former Heisenberg disciple and at the time of writing head of Germany's Max Planck Institute of Physics, explored an explanation in reference to the electromagnetic radiation that surrounds electrons in biomolecules. Consisting of billions of atoms, biomolecules resonate at frequencies between 100 and 1,000 gigahertz. Their longitudinal oscillations are linked to periodic charge displacements, giving rise to the radiation of electromagnetic waves of the same frequency. Dürr speculated that such specifically modulated carrier waves could interlink biomolecules, cells, and even entire organisms, whether they are contiguous or at considerable distance from each other. The process would be similar to superconductivity at very low temperatures, but it could occur at ordinary body temperature in warm-blooded animals.

Dürr concluded that—since according to quantum physics everything is included and incorporated in one indivisible potential reality—it should be possible to find many kinds of connecting links among phenomena. Some of these links may have less the character of a transmission of information between separate things that vibrate at the same frequency (as his own speculations suggest) than the character of a genuinely nonlocal "communion" among seemingly separate but in reality deeply entangled particles and atoms, and the things constituted of them.

Biological Fables: The Interconnected Web of Life
As we have already remarked, Darwin postulated a full and complete separation between genome and phenome, the genetic information

coded in the DNA of the organism's cells, and the environmental influences that reach the organism built according to its genetic information. The genome was said to mutate randomly, unaffected by the vicissitudes that befall the phenome.

The idea that random mutations and natural selection are the basic mechanism of evolution was introduced in 1859, a full century before the nature of the hereditary material would be elucidated together with the specific mechanism by which heritable traits are transmitted. The identification of genes made up of strands of DNA came still later, followed by the discovery of the various modalities of mutation and rearrangement in the genome. The structure of genes in multicellular organisms was clarified in the late 1970s, sufficient DNA sequences to enable the analysis of the origin of genes became available in the 1980s, and the mapping of entire genomes began in the 1990s. Nevertheless, the basic mechanism of evolution described by Darwin was maintained unchanged. The "synthetic theory," the modern version of Darwinism, still insists that randomly produced genetic mutations and the chance fit of the mutants to the milieu evolve one species into another by producing new genes and new developmental genetic pathways, coding new and viable organic structures, body parts, and organs.

Yet random rearrangements within the genome are entirely unlikely to produce viable species. The "search space" of possible genetic rearrangements within the genome is so enormous that random processes are likely to take incomparably longer to produce new species than the time that was available for evolution on this planet. The probabilities are made a great deal worse by the consideration that many organisms, and many organs within organisms, are "irreducibly complex." A system is irreducibly complex, said the biologist Michael Behe, if its parts are interrelated in such a way that removing even one part destroys the function of the whole system. To mutate an irreducibly complex system into another viable system, every part has to be kept in a functional relationship with every other part throughout the entire transformation. Missing but a single part at a single step leads to a dead end. How could this level of constant precision be achieved by random piecemeal modifications of the genetic pool?

An isolated genome working through randomly generated mutations is not likely to produce a new and functional mutant. If such a mutant is in fact produced—and produced time and time again in the course of evolution—the mutation of the genome must be precisely correlated with conditions in the organism's environment. This correlation was often suspected, but in the twentieth century it was dismissed as a mysterious form of "pre-adaptation"—the idea that mutants are somehow spontaneously tuned to the conditions a given species finds in its milieu. Yet unless mutations in the genome are in fact precisely tuned to conditions in their milieu, the resulting mutants will not survive: they will be eliminated by natural selection.

How is it, then, that complex mutants have *not* been eliminated—how could the biosphere be populated by millions of species more complex than algae and bacteria? This could be only if mutations in the genome are highly and quasi-instantly responsive to the environing conditions that affect the organism—if genes and environments form an interconnected system. Evidence is now available that this is indeed the case.

The evidence is statistical, and it goes back to the beginnings of life on this planet. The oldest rocks date from about four billion years, while the earliest and already highly complex forms of life (blue-green algae and bacteria) are over three and a half billion years old. Because even the simplest forms of life manifest a staggering complexity, if the existing species had relied on chance mutations alone, this level of complexity is not likely to have emerged within the relatively short period of about 500 million years. After all, the assembly of a primitive self-replicating prokaryote (primitive nonnucleated cell) is already a complex undertaking. It involves building a double helix of DNA consisting of some 100,000 nucleotides, with each nucleotide containing an exact arrangement of thirty to fifty atoms, together with a bilayered skin and the proteins that enable the cell to take in food. This construction requires an entire series of reactions, finely coordinated with each other.

It is not enough for genetic mutations to produce one or a few positive changes in a species; they must produce the full set. The evolution of feathers, for example, does not produce a reptile that can fly: radical changes in musculature and bone structure are also required, along with a faster metabolism to power sustained flight. Each innovation by itself is not likely to offer evolutionary advantage; on the contrary, it is likely to make an organism less fit than the standard form from which it departed. And if so, it would soon be eliminated by the pitiless mechanisms of natural selection. The cosmologist and mathematical physicist Fred Hoyle has pointed out that life evolving purely by chance is about as likely as a hurricane blowing through a scrap yard assembling a working airplane.

THE BLIND MAN AND THE RUBIK'S CUBE

Fred Hoyle gave a striking example to show why a random selection among even a modest number of alternatives is likely to take far too long to produce usable results. Assume that a blind man is trying to order the scrambled colored faces of a Rubik's Cube. (This is a cube of which the six faces are subdivided into three color-coded sections each. The colors can be ordered by twisting the individual segments.) The blind man is handicapped by not knowing whether any twist he gives the cube brings him closer to or farther from his goal of ordering the sections of the cube. He is obliged to work by random trial and error, with the result that his chances of achieving a simultaneous color matching of the six faces of the cube are in the range of 1:1 to $1:5 \times 10^{18}$. If the blind man is to work through all the possible moves at the rate of one move per second, he will need 5×10^{18} seconds. This, however, he could not do, for 5×10^{18} seconds is 126 billion years—almost ten times more than the age of our universe!

The situation changes dramatically if the blind man receives prompting in his efforts. If he receives a correct "yes" or "no" prompt at each move, the laws of probability show that he will unscramble the cube at an average of 120 moves. Working at the rate of one move per second, he will need not 126 billion years to reach his goal, but two minutes.

Already in 1937, the biologist Theodosius Dobzhansky noted that the sudden origin of a new species by gene mutation might be an impossibility in practice. "Races of a species, and to a still greater extent species of a genus," he wrote, "differ from each other in many genes, and usually also in the chromosome structure. A mutation that would catapult a new species into being must, therefore, involve simultaneous changes in many gene loci, and in addition some chromosomal reconstruction. With the known mutation rates the probability of such an event is negligible." Yet Dobzhansky did not give up the Darwinian theory; instead, he assumed that species formation is a slow and gradual process, occurring on a "quasi-geological scale."

However, the assumption of slow and gradual evolution was contradicted in the 1970s by the finding of new fossils: these show that the "missing links" that appear in the fossil record are not due to the incompleteness of the record, but are true jumps in the course of evolution. New species do not arise through a stepwise modification of existing species—they appear almost all at once. This finding prompted Stephen Jay Gould, then of Harvard, and Niles Eldredge, of the American Museum of Natural History, to advance the theory of "punctuated equilibrium." In this macroevolutionary theory, new species arise in a time span of no more than five to ten thousand years. This may seem like a long time to humans, but as Gould and Eldredge point out, "[I]t translates into geological time as an instant."

The genome must be linked in some way with the milieu in which a species lives, for only such linkage can provide that timesaving "prompt" that allows living species to overcome mutational dead ends and evolve into viable new species. Experimental data back up the statistical evidence. As noted in chapter 3, linkages exist between the phenome and the genome, and they can be mechanical, chemical, bio-chemical, or field-transmitted. Electromagnetic and quantum fields act on the supersensitive organism, and they, too, can trigger adaptive mutations in the genome. Quantum fields appear to link all parts of the organism within the whole organism, and they may also link the whole organism with its external environment. The fact is that the organism is amazingly coherent in itself, and is coherently linked with the world around it.

The coherence of the organism with its environment appears to bring back some aspects of Lamarckism, according to which acquired characteristics can be inherited—what befalls the organism in its milieu can be handed down to its offspring. Although the new findings are not a rediscovery of classical Lamarckism (because characteristics acquired by an organism can be handed down to the offspring only through a modification of the genome), they nevertheless have revolutionary impli-cations. Not surprisingly, the scientific fables that attempt to account for them encounter vivid resistance. Only now do the best-conceived fables command serious attention at the cutting edge of biological research, where systemic biology meets quantum physics in the fledgling disci-pline known as quantum biology.

There is now a fable that is entirely logical and highly supported by evidence: it is that the organism is in some essential respects a quantum system. Being a quantum system, it is linked to other organ-isms as well as to its vital environment much as quanta are linked through space and time: through the A-field, the information field of the vacuum.

4. CONSCIOUSNESS RESEARCH

Puzzles of Transpersonal Consciousness: Intimations of the A-field in the Human Mind

Research on consciousness has become fashionable. There are research institutes, university faculties, scientific journals, and entire book series dedicated to its investigation. Quantum brain researchers look into the interaction of consciousness with the physical world, making use of advanced quantum concepts such as nonlocality, entanglement, phase relations, and hyperspace, among others. The attention of investigators in psychoneuroimmunology, psychosomatic medicine, and other forms of biofeedback research centers on the connection between consciousness and bodily processes, while courageous scientists investigating diverse altered states of consciousness examine the effects of dreams, psychedelic substances, trance, and meditative states, on the assumption that these disclose important and otherwise hidden aspects not only of the subject's own consciousness, but also of his or her links with the outside world. Still more far-flung investigations focus on the effect on consciousness of nonconventional forms of energy known traditionally as prana, kundalini, and chi.

The burgeoning branches of consciousness research use diverse methodologies, but they come to remarkably similar conclusions. The common thrust of their findings is that the human mind is not an isolated entity. To use an expression made popular by Gregory Bateson, it is an "ecology." Consciousness is not fully possessed by the individual, but is present throughout society and perhaps humanity as a whole.

Transpersonal connections. The brains/minds of individual human beings appear to be subtly but effectively linked. So-called primitive peoples have long known of such "transpersonal" links. Medicine men and shamans can induce powers of telepathy through solitude, concentration, fasting, chanting, dancing, drumming, or psychedelic herbs. Whole clans are able to remain in touch with each other no matter where their members roam. Australian Aborigines, the anthropologist A. P. Elkin found, are informed of the fate of family and friends even when they are beyond the range of sensory communication with them. A man far from his homeland will announce that his father is dead, or that his wife has given birth, or that there is some trouble in his country. He is so sure of his facts that he is ready to return home at once.

Many tribal peoples, Marlo Morgan noted, are able to receive input from their environment, do something unique in decoding it, and then consciously act almost as if they had developed some tiny celestial receiver through which they receive universal messages. Modern people have lost everyday access to this "celestial receiver" but laboratory experiments show that they have not lost the receiver itself. Under the right conditions, most people can become aware of the vague but meaningful images, intuitions, and feelings that testify that they are "in touch" with other people and with some aspects of the environment, even when they are beyond the reach of eye, ear, palate, smell, and touch.

Transpersonal contact between individuals has been reported by various psychology and parapsychology laboratories. Thought and image-transference experiments have involved distances between sender and receiver ranging from half a mile to several thousand miles. Regardless of where they have been carried out and by whom, the success rate has been considerably above random probability. The receivers usually report a preliminary impression as a gentle and fleeting form. This form gradually evolves into a more integrated image. The image itself is experienced as a surprise, both because it is clear and because it is clearly elsewhere.

Beyond thought and image transference, a related and apparently likewise universal transpersonal ability is to synchronize the electrical activity of one's brain with the brain of others. A series of experiments carried out by the Italian physician and brain researcher Nitamo Montecucco and witnessed by this writer showed that in deep meditation, the left and right hemispheres of the brain manifest identical wave patterns. Still more remarkably, the left and right hemispheres of *different* subjects become synchronized. In one test, eleven out of twelve meditators achieved a remarkable ninety-eight percent synchronization of their full EEG waves in the complete absence of sensory contact among them.

Another experiment carried out in the presence of the writer took place in southern Germany in the spring of 2001. At a seminar attended by about a hundred people, Dr. Günther Haffelder, head of the Institute for Communication and Brainresearch of Stuttgart, measured the EEG patterns of Dr. Maria Sági, a trained psychologist and gifted natural healer, together with that of a young man who volunteered among the participants. The young man remained in the seminar hall while the healer was taken to a separate room. Both the healer and the young man were wired with electrodes, and their EEG patterns were projected onto a large screen in the hall. The healer attempted to diagnose and then heal the health problems experienced by the subject, who sat with closed eyes in a light meditative state. During the time the healer was concentrating on her task, her EEG waves dipped into the deep Delta region (between 0 and 3 Hz per second), with a few sudden eruptions of wave amplitude. This was surprising in itself, because when someone's brain waves descend into the Delta region, he or she is usually asleep. But the healer was fully awake, even if in a deeply relaxed state. Even more surprising was that the test subject exhibited the same Delta-wave pattern—it showed up in his EEG display about two seconds after it appeared in the EEG of the healer. Yet they had no sensory contact with each other.

Transcultural connections. Anthropological as well as laboratory evidence speaks to the reality of transpersonal connection among individuals, but this is not all. Archaeological and historical evidence testifies that such connection also occurs between entire peoples and cultures.

Subtle, spontaneous contact among cultures appears to have been widespread, as evidenced by the artifacts of different civilizations. In widely different locations and at different historical times, ancient cultures developed an array of similar artifacts and buildings. Although each culture added its own embellishments, Aztecs and Etruscans, Zulus and Malays, classical Indians and ancient Chinese built their monuments and fashioned their tools as if following a shared pattern. Giant pyramids were built in ancient Egypt as well as in pre-Columbian America, with remarkable agreement in design. The Acheulian hand ax, a widespread tool of the Stone Age, had a typical almond or tear-shaped design chipped into symmetry on both sides. In Europe this ax was made of flint, in the Middle East of chert, and in Africa of quartzite, shale, or diabase. Its basic form was functional, yet the agreement in the details of its execution in virtually all traditional cultures cannot be explained by the simultaneous discovery of utilitarian solutions to a shared need: trial and error is not likely to have produced such similarity of detail in so many far-flung populations.

Crafts, such as pottery making, took much the same form in all cultures. At this writer's suggestion, the University of Bologna historian Ignazio Masulli made an in-depth study of the pots, urns, and other artifacts produced by indigenous and independently evolving cultures in Europe, as well as in Egypt, Persia, India, and China during the period from the fifth to the second millennia B.C.E. Masulli found striking recurrences in the basic forms and designs but could not come up with a conventional explanation for them. The civilizations lived far apart in space and sometimes also in time, and did not seem to have had conventional forms of contact with each other.

FOUR PIONEERING TRANSPERSONAL EXPERIMENTS

1. Russell Targ and Harold Puthoff, two physicists, undertook one of the first experiments in controlled transpersonal thought and image transference in the early 1970s. They placed the "receiver" in a sealed, opaque, and electrically shielded chamber and the "sender" in another room where he or she was subjected to bright flashes of light at regular intervals. The brain-wave patterns of both sender and receiver were registered on electroencephalograph (EEG) machines. As expected, the sender exhibited the rhythmic brain waves that normally accompany exposure to bright flashes of light. However, after a brief interval, the receiver also began to produce the same patterns, although he or she was not being directly exposed to the flashes and was not receiving ordinary sense-perceivable signals from the sender.

   Targ and Puthoff also conducted experiments on remote viewing. In these tests, distances that precluded any form of sensory communication between them separated sender and receiver. At a site chosen at random, the sender acted as a "beacon" and the receiver tried to pick up what the sender saw. To document their impressions, receivers gave verbal descriptions, sometimes accompanied by sketches. Independent judges found the descriptions of the sketches matched the characteristics of the site that was actually seen by the sender on average sixty-six percent of the time.

2. In another experiment, in 1994, two physicists, Peter Stewart and Michael Brown, in England, joined with Helen Stewart, a university administrator in New York, to test the telepathic procedure suggested by "Seth" and recounted by Jane Roberts in her best-selling books. Communication was attempted across the Atlantic in fourteen accurately timed sessions between April and September of that year. Detailed records of the observations and impressions were made after each experience via e-mail, and they were stored on automatically dated and timed disks. Though the remotely viewed images were described in terms of associations rather than exact pictorial reproductions of what was seen by the sender, on the whole they corresponded to it. The picture of a meteor shower, for example, came through as a snowstorm; the image of a tower with a rotating restaurant on top was picked up as a globe on a stand. Static images as well as dynamic sequences of images were received—"still pictures" as well as "moving pictures." The physicists concluded that the validity of the telepathic procedure reported by Jane Roberts is established beyond reasonable doubt.

3. The third series of pioneering experiments is the work of Jacobo Grinberg-Zylberbaum, of the National University of Mexico. He performed more than fifty experiments over five years on spontaneous communication among individual test subjects. He paired his subjects inside soundproof and electromagnetic radiation-proof "Faraday cages" and asked

them to meditate together for twenty minutes. Then he placed them in separate Faraday cages where one subject was stimulated and the other not. The stimulated subject received stimuli at random intervals in such a way that neither he (or she) nor the experimenter knew when they were applied. The subjects who were not stimulated remained relaxed, with eyes closed, instructed to feel the presence of the partner without knowing anything about his or her stimulation.

Typically, a series of one hundred stimuli were applied—such as flashes of light, sounds, and short, intense, but not painful electric shocks to the index and ring fingers of the right hand. The electroencephalograph (EEG) brain-wave records of both subjects were then synchronized and examined for "normal" potentials evoked in the stimulated subject and "transferred" potentials in the non-stimulated one. Transferred potentials were not found in control situations where there was no stimulated subject, when a screen prevented the stimulated subject from perceiving the stimuli (such as light flashes), or when the two subjects did not previously interact. But during experimental situations with stimulated subjects and with prior contact among them, the transferred potentials appeared consistently in about twenty-five percent of the cases. A young couple, deeply in love, furnished a particularly poignant example. Their EEG patterns remained closely synchronized throughout the experiment, testifying that their report of feeling deep oneness was not an illusion.

In a limited way, Grinberg-Zylberbaum could also replicate his results. When one individual exhibited the transferred potentials in one experiment, he or she usually exhibited them in subsequent experiments also. The results did not depend on spatial separation between senders and receivers—the transferred potentials appeared no matter how far or how near they were to each other.

4. The fourth experiment is particularly intriguing: it involves dowsing. It is known that dowsers can often pinpoint the location of water veins with great accuracy. Dowsing rods as well as pendulums respond to the presence of underground water, magnetic fields, and even oil and other natural substances. (Evidently, it is not the dowsing rod itself that responds to the presence of water and other things, but the brain and nervous system of the person who holds the rods, for the rods do not move unless they are held by a dowser. The rods merely make the information visible: they enlarge the subtle and involuntary muscle responses that move the arm of the dowser.) It now appears that dowsers can also pick up information that is not produced by natural causes but is projected long-distance by the mind of another person. "Dowsable" lines, figures, and shapes can be created by the conscious intention of one person, and these lines, figures, and shapes affect the mind and body of distant persons who have not been told what has been created and where. Their rods move just as if the figures, lines, and shapes were due to natural causes immediately in front of them. This is the finding of a series of remote-dowsing experiments carried out in the past ten years by Jeffrey Keen, a renowned engineer, together with colleagues at the Dowsing Research Group of the British Society of Dowsers.

In a considerable number of experiments, the exact shapes created by the experimenter could be identified by the dowsers. It turned out that the shapes could be positioned with an accuracy of a few inches even when created thousands of miles away. The accuracy of positioning was not affected by the distance between the person creating the dowsable fields and the physical location of the fields: the same results were produced whether the experimenter created a dowsable shape a few inches or five thousand miles away. There was no difference whether the experimenter stood on the ground, was in an underground cave, flew in a plane, or was inside a Faraday cage. Time did not seem to be a factor either: the fields were created faster than measurements could be taken, even over large distances. Time also proved irrelevant because the fields remained present and stable at all times after their creation. In one case they endured for more than three years. But they could be canceled if the person who created them wanted it.

Keen concluded that dowsable fields are created and maintained in an "Information Field that pervades the universe." The brain interacts with this field and perceives dowsable fields as holograms. This, according to Keen and the Dowsing Research Group, is an instance of nonlocal interaction between the brain and the field by different and even distant individuals.

NDEs and other altered-state-of-consciousness experiences. There is now a significant body of evidence that the range of information reaching brain and consciousness transcends the range coming through eyes and ears. A remarkable kind of evidence comes from the investigations of Kenneth Ring, a British near-death-experience (NDE) researcher. Not content with finding evidence of veridical out-of-body experiences by ordinary people at the portals of death (well documented ever since Elisabeth Kübler-Ross began researching the NDE phenomenon), Ring investigated such experiences in blind people.

In one series of tests, fifteen out of twenty-one blind people whose physical condition approached death reported fully sighted visions. (Of the remaining six, three were not sure whether they saw or not, and three did not appear to see at all.) Among those who reported sighted visions, Ring found no obvious differences whether they had been blind from birth, had lost sight later in life, or suffered from severe visual impairment. Furthermore, the experiences they reported were much the same as those reported by sighted people. Ring tried to explain these findings by the usual skeptical arguments, such as: that they are only apparent and not actual; that they are similar to dreams; that they constitute retrospective reconstructions of prior experiences; and that they are a form of "blindsight" due to receptors in the brain or on the skin. However, he found that no such explanation can account for the clear visual features of the experiences, or indeed for the finding that many of their features were subsequently confirmed as veridical perception.

NDEs occur in altered states of consciousness, as do other forms of out-of-body experiences. Meditation, intense prayer, fasting, rhythmic movements, and controlled breathing also produce altered states, and all these states prove conducive to the reception of nonsensory information. When consciousness is in an altered state, the brain seems to function in a mode in which information that does not fit the commonsense conception of the world is not repressed. By contrast, ordinary waking consciousness is a strict censor: most people have been "brainwashed" to filter out all experiences not clearly and evidently conveyed by eyes and ears. Parents tell their children not to imagine things, teachers insist that they should stop dreaming and be sensible, and peer groups, already brainwashed, laugh at the child who persists. As a result, modern youngsters grow up to be commonsense individuals for whom everything that does not accord with the dominant materialist idea of the world is denied and repressed. In altered states of consciousness, however, strange items enter consciousness. And not everything that enters turns out to be imaginary . . .

The ability of altered states to convey veridical information about the world was known to traditional peoples who prized and cultivated them for the power they confer. But modern people think of altered states as pathological—a sign of disease, of dementia, or of being high on drugs. Only dreaming, daydreaming, alcoholic intoxication, and sexual orgasm are considered "normal" deviations from waking consciousness. Natural healers, leading-edge psychiatrists, and consciousness researchers have a different view of such states. The psychiatrist John Nelson, for example, considers altered states basic to the human psyche, with one end of the spectrum shading into madness and the other reaching to the loftiest realms of creativity, insight, and genius.

In over forty years of clinical experience, the renowned psychiatrist Stanislav Grof investigated the power of altered states. Gathered in more than four thousand psychedelic sessions with various mind-altering substances, two thousand sessions conducted by colleagues, and over thirty thousand sessions using the holotropic breathing method, Grof's experience indicates that when the censorship of the waking consciousness is not operative, information can reach the mind from almost any part or aspect of the universe.

For example, in the "experience of dual unity," a person in an altered state of consciousness can experience a loosening and melting of the boundaries of the body ego and a sense of merging with another person in a state of unity and oneness. In this experience, despite the feeling of being fused with another, one retains an awareness of one's own identity. In the experience of "identification with other persons," an individual, while merging experientially with someone else, can experience complete identification to the point of losing the awareness of his or her own identity. Identification is total and complex, involving body image, physical sensations, emotional reactions and attitudes, thought processes, memories, facial expressions, typical gestures and mannerisms, postures, movement, and even the inflection of the voice. The person with whom the given individual identifies can be someone in his or her presence or it can be a distant person, alive now or long since dead.

In "group identification and group consciousness" there is a further extension of consciousness and melting of ego boundaries. Rather than identifying with an individual, a person can have a sense of becoming an entire group of people with some shared racial, cultural, national, ideological, political, or professional characteristic. Identification can focus on a social or political group, the people of an entire country or continent, all members of a race, or all believers of a religion. The depth, scope, and intensity of this experience can reach extraordinary proportions: some people experience the totality of suffering of all the soldiers who have ever died on the battlefield since the beginning of history; the desire of revolutionaries of all ages to overthrow a tyrant; or the love, tenderness, and dedication of all mothers in regard to their babies.

Telesomatic connections. Transpersonal and transcultural effects are not limited to contact and communication between the minds of different, and possibly distant, people: effects can be transmitted also from the mind of one person to the *body* of another. This "telesomatic" effect was likewise known to traditional cultures; anthropologists called it "sympathetic magic." At the University of Nevada, the experimental parapsychologist Dean Radin tested it under controlled laboratory conditions.

In Radin's experiments, the subjects created a small doll in their own image and provided various objects (pictures, jewelry, an autobiography, and personally meaningful tokens) to "represent" them. They also gave a list of what makes them feel nurtured and comfortable. These items and the accompanying information were used by the "healer"—who functioned analogously to the "sender" in thought- and image-transfer experiments—to create a sympathetic connection to the "patient." The latter was wired up to monitor the activity of his or her autonomous nervous system (electrodermal activity, heart rate, and blood pulse volume) while the healer was in an acoustically and electromagnetically shielded room in an adjacent building. The healer placed the doll and other small objects on the table in front of him and concentrated on them while sending randomly sequenced "nurturing" (active healing) and "rest" messages.

The electrodermal activity of the patients, together with their heart rate, was significantly different during the active nurturing periods than during the rest periods, while blood pulse volume was significant for a few seconds during the nurturing period. Both heart rate and blood flow indicated a "relaxation response"—which makes sense since the healer was attempting to "nurture" the subject via the doll. On the other hand, a higher rate of electrodermal activity showed that the patients' autonomic nervous system was becoming aroused. Why this should be so was puzzling, until the experimenters realized that the healers nurtured the patients by rubbing the shoulders or stroking the hair and face of the dolls that represented them. This, apparently, had the effect of a "remote massage" on the skin of the patients!

Radin and colleagues concluded that the local actions and thoughts of the healer are mimicked in the distant patient almost as if healer and patient were next to each other. Distance between sender and receiver seems to make no difference. This was confirmed in a large number of trials conducted by the experimental parapsychologists William Braud and Marilyn Schlitz regarding the impact of the mental imagery of senders on the physiology of receivers. Braud and Schlitz found that the mental images of the sender could reach out over space to cause changes in the physiology of the distant receiver. The effects are comparable to those that one's own mental processes produce on one's body. "Telesomatic" action by a distant person is similar to, and nearly as effective as, "psychosomatic" action by individuals on themselves.

Distant mental effect can be produced on other forms of life as well. In a series of experiments, the lie-detector expert Cleve Backster attached the electrodes of his lie detector to the leaves of a plant in his New York office. He recorded the changes in electric potentials on the surface of a leaf just as he would record such changes in a human subject. To his amazement, Backster found that the plant registered his own emotions—showing sudden jumps and wild fluctuations at the precise moment when Backster himself had a strong emotional reaction, whether he was in the office or away from it. Somehow, the plant seemed to "read" his mind. Backster speculated that plants have a "primary perception" of the people and events around them. Subsequently he tested many varieties of plants, cells, and even animals and found the same kind of response in the lie detector. The leaves of plants responded even when they were ground up and the remains distributed over the surface of the electrodes!

Subsequently, Backster undertook a series of experiments in which he tested white cells (leukocytes) taken from the mouths of his test subjects. The procedure of obtaining the cells has been perfected for purposes of dentistry and produces a pure cell culture in a test tube. Backster moved the culture to a distant location, anywhere from five meters to twelve kilometers from his subjects. He placed the electrodes of the lie detector on the distant culture and provoked some emotion-producing response in his subjects. In one case he had a young man look at an issue of *Playboy* magazine. Nothing spectacular occurred until the young man came to the centerfold and saw a photo of actress Bo Derek in the nude. At that moment the needle of the lie detector attached to the distant cell culture began to swing, and kept fluctuating as long as the subject was looking at the picture. When he closed the magazine, the needle returned to trace a normal pattern, but was suddenly reactivated when the young man decided to have another look at the magazine.

In another test a former Navy gunner who was at Pearl Harbor during the Japanese attack was shown a TV program depicting the attack. He showed no particular reaction until the face of a Navy gunner appeared on the screen, followed by a shot of a Japanese plane falling into the sea. At that moment the needle of the lie detector attached to his cells twelve kilometers away jumped. Subsequently, both he and the young man confirmed that they had a strong emotional reaction at these particular points. It made no difference whether the cells were a few meters or several kilometers away. The lie detector displayed exactly the response it would have displayed if it had been attached directly to the subject's body. Backster was forced to conclude that a form of "biocommunication" is taking place for which we have no adequate explanation.

Psi-Fables: Nonlocal Consciousness

Psi-fables abound in the world: they are the woof and warp of the popular esoteric schools. Such fables are now also produced by scientists, if only by a handful of insightful and courageous ones. A few examples stand out.

William James, known as the father of American psychology, declared: "Out of my experience . . . one fixed conclusion dogmatically emerges . . . that we with our lives are like islands in the sea, or like trees in the forest. The maple and pine may whisper to each other with their leaves. . . . But the trees also commingle their roots in the darkness underground, and the islands hang together through the ocean's bottom. Just so there is a continuum of cosmic consciousness, against which our individuality builds but accidental fences, and into which our several minds plunge as into a mother sea or reservoir. . . ." The physicist Erwin Schrödinger echoed a similar insight. "In all the world," he wrote, "there is no kind of framework within which we can find consciousness in the plural; this is simply something we construct because of the spatio-temporal plurality of individuals, but it is a false construction . . . the self-consciousness of the individual members are numerically identical both with [each] other and with that Self which they may be said to form at a higher level."

David Bohm came to essentially the same conclusion: "Deep down the consciousness of mankind is one," he affirmed. "This is a virtual certainty because even in the vacuum matter is one; and if we don't see this it's because we are blinding ourselves to it." In 2001 Henry Stapp placed the psi-fable of nonlocal consciousness in the current physics framework. "The new physics," he remarked, "presents prima facie evidence that our human thoughts are linked to nature by nonlocal connections: what a person chooses to do in one region seems immediately to affect what is true elsewhere in the universe. This nonlocal aspect can be understood by conceiving the universe to be not a collection of tiny bits of matter, but rather a growing compendium of 'bits of information.' "

Unless scientists are well established, psi-fables are dangerous territory for them, making them a target for criticism and even censure. But the accumulation of evidence regarding nonlocal connection between the brains and minds of people is now so significant that even the mainstream science community is taking notice. In the spring of 2000, a collection of papers published by the ordinarily conservative American Psychological Association reviewed and assessed the relevant evidence. Edited by Etzel Cardeña, Steven Jay Lynn, and Stanley Krippner, *Varieties of Anomalous Experience: Examining the Scientific Evidence* reviewed hallucinatory experiences, synesthesia, lucid dreaming, out-of-body experiences, psi-related experiences, alien abduction experiences, past-life experiences, near-death experiences, anomalous healing experiences, and mystical experiences. The authors agreed that these experiences could not be dismissed either as illusory or as signs of psychopathology. They are more widespread than has been generally assumed, and have a real impact on the people who experience them. Yet, they said, there are no definitive explanations for them.

This conclusion is characteristic of the state of the art in academic consciousness research. The nonlocal aspects of consciousness are less and less disputed, but not significantly better understood. As the pioneer altered-state researcher Russell Targ put it, "[I]t is all phenomena." Since meaningful explanation is beyond the bounds of legitimate research, the investigation of the phenomena is shifted into the domain of "para"-psychology. But at least under that heading, the academic community is taking an interest. Utrecht University, in the Netherlands, and Edinburgh University, in Scotland, have chairs in parapsychology and, as of 2004, Sweden's Lund University also has a chair for "parapsychology, hypnology, and clairvoyance."

Recognition that there is a bona fide scientific explanation for the observed nonlocality of consciousness would give legitimacy to research on psi phenomena and open the way to a better understanding of the as yet mysterious dimensions of the human mind. Such explanation is now at hand. The information field that links quanta and galaxies in the physical universe and cells and organisms in the biosphere also links the brains and minds of humans in the sociosphere. This A-field creates the human information pool that Carl Jung called the collective unconscious and Teilhard de Chardin the noosphere—and that scientists such as Erwin Schrödinger, David Bohm, William James, and Henry Stapp have not hesitated to discuss and to affirm.

SIX

The "A-Field Effect"

Let us take stock. We have a set of puzzles before us, and a variety of fables that seek to explain them. Even though they are couched in different conceptual and theoretical frameworks, the fables point to a shared fundamental conclusion: things in the real world are not entirely separate from each other. Not only are they linked by flows of energy; but they are also linked by flows of information. How does this linking actually occur? This is the question regarding the effect of the A-field on the world. It is the question to which we now turn.

THE VARIETIES OF A-FIELD EFFECT

That the A-field informs all things with all other things follows as the simplest and most meaningful explanation of the nonlocality and entanglement we have encountered in physics and in cosmology, as well as in biology and in consciousness research. But in itself this is a "fable," even if a plausible one, and not (or not yet) a scientific theory. We also need to explain how the A-field *works*.

Exploring the working of the A-field is not a simple matter, for the A-field cannot be perceived. This field is not an imaginary phantasm, however, for it produces an effect and its effect can be perceived. This is the rule and not the exception in regard to the other fields postulated in science. For example, the gravitational field cannot be perceived: when we drop an object to the ground, we see the object falling but not the field that makes it fall—we see the effect of the "G-field" but not the G-field itself. The effect of the G-field is gravitation among separate masses; the effect of the EM field is electrical/magnetic wave transmission; and the effect of the weak and strong nuclear fields is attraction or repulsion among masses at extreme proximity to each other. What, precisely, is the effect of the A-field?

We claim to know what it is that produces the A-field effect: it is the quantum vacuum. The question is how the A-field of the vacuum affects the particles and the more complex things that are integrated ensembles of particles—atoms and molecules, cells and organisms, and stars and galaxies—that exist in space and time.

The A-field conveys information, and this information, subtle as it is, has a notable effect: it makes for correlation and creates coherence. This "in-forming" of everything by everything else is universal, but it is not universally the same. *Universal* information does not mean *uniform* information. The A-field conveys the most direct, intense, and therefore evident information between things that are closely similar to one another (i.e., that are "isomorphic"—have the same basic form). This is because A-field information is carried by superposed vacuum wave–interference patterns that are equivalent to holograms. We know that in a hologram every element meshes with isomorphic elements: with those that are similar to it. Scientists call such meshing "conjugation"—a holographic pattern is *conjugate* with similar patterns in any assortment of patterns, however vast.

Practical experience bears this out. Using the conjugate pattern as the "key," we can pick out any single pattern in the complex wave pattern of a hologram. We need merely to insert the given wave pattern into the welter of patterns in the hologram and it attaches to the patterns that are conjugate with it. This is similar to the phenomenon of resonance. Tuning forks and strings on musical instruments resonate with other forks and strings that are tuned to the same frequency, or to entire octaves higher or lower than their frequency, but not with forks and strings tuned to different frequencies. Such a selective response is also encountered on the Internet. When we input the code of the Web site we are looking for, the system matches it with the code that corresponds to a given site. It unlocks the one door we want among the millions of doors on the Web.

When we apply the conjugate principle to the interference patterns in the A-field, we get a simple and logical picture. Through the holograms created in and conveyed by the A-field, things are directly "in-formed" by the things that are most like them. For example, an amoeba is directly informed by other amoebas. This does not mean that things that are unlike one another would not be mutually informed. They are so informed, but the informational effect is not equally evident in all cases. Amoebas are informed by other single-celled organisms, and they are also informed by far simpler entities such as molecules and by far more complex ones such as multicellular organisms. But information by things on other levels is less intense and evident than information by things that correspond to a thing's own level. The same goes for human beings. We are directly informed by fellow humans, yet we are also informed, though less directly, by animals, plants, and all of nature. Information conveyed through the A-field subtly tunes all things to all other things and accounts for the coherence we find in the cosmos, as well as in living nature.

The A-Field Effect in the Cosmos

As we have seen in chapter 5, through torsion waves in the vacuum the A-field links things and events in the universe at staggering speeds—a billion times the velocity of light. The interference patterns of torsion waves create cosmic-scale holograms, the holograms of stars and entire stellar systems. These extend throughout our universe and correlate its galaxies and other macrostructures.

The torsion-wave interference patterns of stars and stellar systems create the hologram of the galaxy, and the interfering vacuum torsion waves of the metagalaxy (the set of all galaxies) create the hologram of the universe. The hologram of the universe is conjugate with the hologram of the galaxies, and thus this encompassing hologram creates coherence among the galaxies—it correlates the paths of their evolution. This A-field effect is extremely subtle yet it is effective: stars and galaxies evolve coherently throughout the universe, even across distances that could not have been traversed by any light or any signal known to modern physics.

The "fine-tuning of the universal constants"—why the basic parameters of the universe are so astonishingly coordinated that complex systems such as ourselves can arise in it—is likewise an effect of the A-field. We know that the Big Bang was incredibly precise in regard to its parameters, and that the energy density of the vacuum was precisely such that the particles created in that explosion did not fly apart before they could condense into galaxies and stars and a variety of potentially life-harboring planets. In a less finely tuned universe we would not be here to marvel at this precision. With only a minuscule deviation (such as one part in a billion in the value of a universal force such as electromagnetism or gravitation, or a tiny excess in energy density), the universe would have been incapable of producing conditions where living organisms can emerge and evolve.

In the Big Bang theory the fine-tuning of the constants has no convincing explanation: mainstream cosmology can only assume that the pre-space of the universe was random, with chance fluctuations in the vacuum. However, it is extremely unlikely that chance fluctuations would have resulted in precisely those fluctuation patterns that could give rise to a finely tuned universe such as ours.

String theorists do offer an explanation for the fine-tuning of our universe. Leonard Susskind, for example, suggests that the energy density of the vacuum varies from region to region. There are so many locally different "vacua"—perhaps of the order of 10^{500}—that we can be reasonably certain of finding at least one that has the properties we are looking for. Since we are here to search for it, we have evidently found it: it is our particular "local vacuum"—our region of the cosmos.

There is, however, a simpler explanation. The "Bang" that gave birth to our universe, and the vacuum in which it occurred, was informed by a prior universe—a previous cycle of the Metaverse. Whether the universe is infinite or finite in space (and that is still not clear at present), it is most likely not finite in time: the cosmos is not limited to a single-cycle universe. This is relevant to our query, for it is evident that all universes that exist, and have ever existed, arise in the quantum vacuum. The particles that make up a given universe spring from the vacuum and fall back into it at the end of the universe's life cycle. The vacuum existed before any universe was born, and will continue to exist after all matter in that universe vanishes into black holes. In the course of a vast series of universes, the vacuum becomes progressively tuned to the processes that take place in the universes that succeed one another.

The A-field effect gives us a simple and logical explanation of the fine-tuned features of our own universe. When this universe was born, the Bang that created it and the vacuum in which that event took place were not randomly configured. They were informed by prior universes, much like at conception the genetic code of a zygote is informed by the genetic code of the parents. That this was the case is far more plausible than a random selection from among an astronomically large number of disconnected universes—or among a similarly mind-boggling number of "vacua" in one and the same universe.

The A-Field Effect in the Living World

There is an A-field effect throughout nature—the A-field affects also organisms. This stands to reason. In the domains of life, the individual holograms of the molecules and cells of an organism mesh ("conjugate") with the encompassing hologram of the whole organism. In consequence, there is a subtle yet effective correlation among the organism's molecules, cells, and organs, producing nearly instantaneous coherence within the organism. This conjugate relationship exists whether the molecules and cells are next to each other or distant. As we have seen, experiments show that cells that once belonged to an organism remain connected with that organism even when they are miles away from it.

Information through the A-field accounts not only for the quasi-instant coherence of all parts of an organism, but also for the subtle but effective correlation between organisms and environments. The holograms of entire colonies, groups, and communities of organisms are conjugate with the hologram of the ecology of which they are a part. The hologram of the ecology in which organisms are embedded correlates all organisms in that ecology, down to the structure of their genome. Thereby, the ongoing variation of the genome is subtly informed, increasing the probabilities that when the environment changes, the genome comes up with mutations that are viable in the new milieu.

The same principle accounts for the astonishingly rapid evolution of life in the primeval seas of the young Earth. We have seen that the oldest rocks date from about four billion years ago, while the earliest and already highly complex forms of life—blue-green algae and bacteria—are over three and a half billion years old. Creating these life-forms required a coordinated and complex series of reactions, where missing but a single step would have led to a dead end. A random mixing of the "molecular soup" in the shallow primeval seas is unlikely to have accomplished this feat in the available time span. But the mixing of the molecules on the surface of the primeval Earth was not purely random: it was informed by the traces of already evolved life! Evidently, these traces were not those of life on Earth, since we are speaking of the earliest beginnings of biological evolution on this planet. They were the traces of life on other planets.

The "informational seeding" of biological evolution on Earth is entirely plausible. The zero-point field of the vacuum extends throughout the universe and can carry the torsion-wave interference patterns of particles and particle systems throughout space. Wherever the vacuum's holograms penetrate, they bring with them information on the life-forms that evolved in that region of the universe. Since in our galaxy life is likely to have evolved on other planets prior to its evolution on Earth, the holographic traces of other biospheres must have been present in the vacuum at the time the first forms of life appeared on this planet. These traces must have been sufficiently conjugate with the life-forms that emerged on the young Earth to produce a subtle yet decisive effect on them. This accelerated the trials and errors of evolution, increasing the chances that the turbulent mixing of the molecular soup would hit on stable, self-maintaining combinations.

Life on Earth was informed by life elsewhere in the universe—just as Earth life now informs life on other life-bearing planets, wherever they may exist in this galaxy and beyond.

IN CONCLUSION . . .

Beyond the puzzle-filled world of the mainstream sciences, a new concept of the universe is emerging. The established concept is transcended; in its place comes a new/old concept: the informed universe, rooted in the rediscovery of ancient tradition's Akashic Field as the vacuum-based holofield. In this concept the universe is a highly integrated, coherent system, much like a living organism. Its crucial feature is information that is generated, conserved, and conveyed by and among all its parts. This feature is entirely fundamental. It transforms a universe that is blindly groping its way from one phase of its evolution to the next into a strongly interconnected system that builds on the information it has already generated.

A cosmic field that underlies and links all things in the world is a perennial intuition, present in traditional cosmologies and metaphysics. The ancients knew that space is not empty: it is the origin and the memory of all things that exist and have ever existed. But this knowledge was based on philosophical or mystical insight, the fruit of private and unrepeatable if often seemingly indubitable experience. The current rediscovery of the Akashic Field reinforces qualitative human experience with quantitative data generated by science's experimental method. The combination of unique personal insight and interpersonally observable and repeatable experience gives us the best assurance we can have that we are on the right track: that a cosmic information field connects organisms and minds in the biosphere, and particles, stars, and galaxies throughout the cosmos.

Nature's information field is now being rediscovered at the cutting edge of the sciences. It emerges as a powerful fable and—as sustained research deepens and specifies the theory of the A-field—as a main pillar of the scientific world picture of the twenty-first century. This will profoundly change our concept of ourselves and of the world.

The rediscovery of the A-field will also change our world itself. When people realize that the age-old intuition that space does not separate things but links them has a bona fide scientific explanation, the genius for innovation inherent in modern civilization will find ways to make practical use of it. As people learn to work with the A-field, untold ways will come to light for beaming active and effective "in-formation" from one place to another, instantly and without the expenditure of energy. This will not only enable quantum computation, but also pave the way to an entire series of technological breakthroughs. We will learn to teleport not just bits of quanta but atoms and molecules, living cells and organs, as well as aspects and elements of consciousness.

The possibilities can hardly be fully grasped today. But we should not be surprised at being surprised—over and over again.

PART TWO

———

EXPLORING THE
INFORMED UNIVERSE

INTRODUCING THE INFORMED UNIVERSE

The informed universe is a universe where the A-field is a real and significant element. Thanks to this field, this universe is of mind-boggling coherence. All that happens in one place happens also in other places; all that happened at one time happens also at all times after that. Nothing is "local," limited to where and when it is happening. All things are global, indeed cosmic, for the memory of all things extends to all places and all times. This is the concept of the informed universe, the view of the world that will hallmark science and society in the coming decades.

The informed universe is not a universe of separate things and events, of external spectators and an impersonal spectacle. Unlike the world of the mainstream sciences, it is not even materialistic. Matter—the kind of "stuff" that is made up of particles joined in nuclei joined in atoms joined in molecules that may also be joined in cells joined in tissues joined in organisms joined in ecologies—is not a distinct reality. It is energy bound in quantized wave-packets. The classical idea, that all there is in the world is matter, and that all matter was created in the Big Bang and will disappear either in black holes or in a Big Crunch, was a colossal mistake. And the belief that when we know how matter behaves we know everything—a belief shared by classical physics and Marxist ideology—was a colossal pretense. Such views have been definitively superseded. There are more things in this universe than classical scientists, engineers, and Marxists have ever thought of! And many of the things that are in this world are more amazing than writers of science fiction could ever imagine.

But the truly remarkable feature of the informed universe is not that matter is not its fundamental feature. What is remarkable is that everything that happens in it affects—"informs"— everything else. This is not as strange as it might seem: we have noted in chapter 4 that even in the sea around us every thing affects every other thing. An even more familiar example is the aquarium many of us had as children. Christopher Laszlo, the author's son, had such a fish tank in his teens and he maintains that understanding what happens in it is a good way to understand what happens in the universe at large.

THE FISH TANK AND THE INFORMED UNIVERSE: A SUGGESTIVE METAPHOR

Contributed by Christopher Laszlo*

Imagine you are standing in front of a huge panoramic-view aquarium. Angelfish and dwarf cichlids hover delicately while giant gourami and red-striped tiger barbs chase a few lazy scavenger fish on the pebbled floor. Silver neons flash among the African water ferns and Amazon sword plants. Small bubbles of air rise to the drone of an electric filter.

Suddenly two motorized toy submarines are introduced from the water's surface and sink halfway down. The fish streak around the aquarium walls a few times and then settle down as the apparent danger vanishes.

Now look closely at the motion of the submarines. They weave and bob with the movement of the fish, even with the rising bubbles of air. When the submarines are switched on, they glide through the water, creating little underwater wakes that draw in the fish and cause the plants to sway. At times a submarine pulls a fish into its wake, and the fish, in reacting to this movement, struggles to get away and creates turbulence that causes the submarine to veer precipitously on its side.

---

*Christopher Laszlo is senior partner of Sustainable Value Partners, a management consulting firm, and coauthor of *The Insight Edge* (with Ervin Laszlo) and *Large-Scale Organizational Change* (with Jean-François Laugel). His latest book is *The Sustainable Company*. Christopher Laszlo lives in Great Falls, Virginia.

Every motion has an impact on everything else in the tank. Every fish, plant, submarine, pebble, and bubble is connected by motion through the water in the form of waves. Although you cannot see it, the intersecting waves carry information about the things that created them. The wake of the submarine propeller codes a different set of data than the ripples of a dorsal fin. As the two waves collide, the submarine and the fish mutually influence each other, conveying each other's location, speed, and size.

You are looking at a simple model of the universe according to the theory of the A-field. In this theory the underlying physical reality is a holographic field in which every thing—be it a particle, an atom, a molecule, an amoeba, a mouse, or a human being—is connected with every other thing. And every thing affects every other thing through wave pressures that literally shape the things around them.

There are a few important differences between the fish tank model and the A-field-informed universe. In the fish tank, the waves contain information as well as a physical force—you can feel the impact of a wave under water. In the A-field, the waves carry information without carrying force, meaning that you can't feel them. In the fish tank, the waves eventually slow down and disappear. In the A-field, the waves never attenuate because they are moving through a frictionless medium, with nothing to slow their progress. These first two differences between the fish tank and the universe arise because the A-field is a medium that, like the supercooled helium used in superconductivity experiments, cannot be registered by conventional means. You can't see or feel waves in the A-field. Energy moves through superconductive material without ever slowing down or diminishing, unlike electric impulses moving through copper—which is why phone lines need repeaters to carry signals over long distances. In the medium of the A-field, things move effortlessly without encountering any observable resistance. For this reason, leading scientists in the past concluded that space is merely a void. Sir Isaac Newton himself believed that the vacuum of space is a passive receptacle through which physical objects move, obeying the laws of motion he discovered.

But wait—the informed universe becomes ever more strange. In the fish tank, waves travel at relatively earthbound speeds of up to a few hundred miles an hour over tiny distances. In the A-field, waves can travel faster than the speed of light—faster than 186,000 miles per second! This very high speed of information transmission accounts for events that appear to be synchronized over great distances—a kind of instant correlation, known as nonlocality, that scientists are discovering in a number of disciplines. Think how instantly every molecule in your body adjusts to the thousands of biochemical reactions produced every second, or how a thought that suddenly entered your mind also entered your loved one's mind at precisely the same moment, even though he or she was hundreds of miles away at the time.

In the fish tank, "what you see is what you get": a tiger barb is the same color and shape every time you look at it. In the informed universe, each of the tiniest building blocks of physical reality (known by strange names such as quarks, gluons, and bosons) exists as a potential of many different states. Their potentiality is said to collapse into an actual state when observed or otherwise interacted with. It's as if a tiger barb fish "potential" existed that, when observed, became one of several possible actual tiger barbs—sometimes silver and thin, sometimes striped and fat, other times transparent.

The A-field ties together all physical systems in a highly coherent whole. This means that pure chance, the roll of the die, plays no fundamental role in evolution, unlike Darwin's theory of random mutations that lead to the survival of the fittest. The

A-field continuously interacts with matter at every level from subatomic to cosmic to influence the way every living thing grows, adapts, and evolves. This makes for a highly coherent world in which things at one level (such as atoms) are influenced by things at another level (such as human beings), which in turn are influenced by still other levels all the way up to the universe itself—and even prior universes, helping to account for the finely tuned coherence of our own universe as we know it.

In this perspective the cosmos is intrinsically creative, preserving and renewing the imprint of all that exists. The A-field is a kind of active memory field encompassing space (it is everywhere) and time (it endures). It is as if all the fish and plants in the fish tank were physical manifestations of the water, interconnected by the water in such a way that whatever happens to one influences what happens to all others in a mutually dependent system, evolving together in a delicate dance of all life and all of nature.

NEWTON	DARWIN	FREUD	EINSTEIN	LASZLO
clockwork mechanism	survival of the fittest	subconscious self-centered	relativity of space-time	coherent fine-tuned interconnected whole

SEVEN

The Origins and Destiny of Life and the Universe

WHERE EVERYTHING CAME FROM— AND WHERE IT IS GOING

In the chapters that follow we ask questions that concern the nature and future of the universe and the principal kind of things in it. Where did everything come from—and where is it going? Is there life elsewhere in this galaxy and beyond? And if there is, will it evolve to higher stages or dimensions?

We also ask about the nature of consciousness. Did it originate with *Homo sapiens*, or is it part of the fundamental fabric of the cosmos? Will it evolve further in the course of time—and what kind of impact will it have on us and on our children when it does?

Then we probe still deeper. Does consciousness cease at the physical death of the body or does it continue to exist in some way, in this or in another sphere of reality? And, last but not least, could the universe itself possess some form of consciousness, a cosmic or divine root from which our consciousness has grown, and with which it remains subtly connected?

We begin with perhaps the greatest of the "great questions." *Where did the universe come from?*

People have never ceased to wonder about the origins and the destiny of the world. The earliest answers were couched in the mystical worldview, followed by the worldviews of the great religions. In regard

121

to concepts of origin and destiny, the classical views of East and West were remarkably consistent: they both envisaged the origins of the universe as a stupendous process of self-creation. But with the rise of monotheistic religion in the West, the creation story of the Old Testament replaced mystical and metaphysical accounts. Throughout the Middle Ages, Christians, Muslims, and Jews believed that an all-powerful God created the sky above and the Earth below, and all things in between, with purpose and intent, just the way we find them.

In the nineteenth century, the Judeo-Christian account of creation came into conflict with the theories of modern science, in particular with Darwinian biology. A vivid contrast arose between the view that everything we behold was created intentionally by a divine power and the concept according to which living species evolve on their own, from simpler common origins. The contrast fueled endless debates, surviving to this day in the controversy surrounding the teaching of "creationist" vs. "evolutionist" theories in public schools.

Since the 1930s, the Judeo-Christian creation story has had to contend not only with the Darwinian doctrine of biological evolution, but also with physical cosmology. Newton's clockwork universe required a Prime Mover to wind it up and get it going, and this could be attributed to the work of a Creator. Subsequently Einstein's steady-state universe could do without a Creator, for it persisted from the beginnings of time the same as it is today. But when the steady-state universe was replaced by the Big Bang theory's explosively expanding universe, questions arose again about the world's origins. If the universe was born in a Big Bang 13.7 billion years ago and will end either in the Big Crunch some two thousand billion years in the future or in the evaporation of the last galactic-cluster-sized black holes at the almost inconceivable time horizon of 10^{122} years, the question that comes to mind is: *What was there before all this began—and what will be there after it is over?*

The best the "BB theory" can say about how the universe came into being is that a random instability took place in a fluctuating cosmic vacuum, the pre-space of the universe. It cannot say either *why* this instability occurred or why it occurred *when* it occurred. And otherwise than through implausibly speculative fables—such as a cosmic roulette among a large number of randomly created universes—it also cannot say why the universe came to be the *way* it came to be: why it has the remarkable properties it now exhibits. The question returns, it seems, to the domain of religion and mysticism. But giving up on science would be premature. The Big Bang theory is not the final word; the new cosmologies have more to say about cosmic origins.

As we have seen, there are sophisticated cosmologies that tell us that our universe is not the only universe. There is also a meta-universe or Metaverse that was not created in the Bang that created our universe (which was but one of many explosions, so it no longer qualifies for the adjective "Big"); nor will the Metaverse itself come to an end when all the matter created by this particular Bang vanishes in the collapse of the last black holes. The insight that dawns is that *the* universe existed prior to the birth of *our* universe, and it will continue to exist after our universe's demise. *The* universe is the Metaverse, the mother of our universe and perhaps of myriad other universes.

Cosmologies of the Metaverse are in a better position than the Big Bang theory (which is limited to *our* universe) to speak of conditions that reigned before, and will reign after, the life cycle of our universe. The quantum vacuum, the subtle energy and information sea that underlies all matter in the universe, did not originate with the Bang that produced our universe, and it will not vanish when matter created by that explosion dies back. The subtle energies and information that underlie this universe were there before its particles of matter appeared and will be there after these particles disappear. Thus, the deeper reality is the quantum vacuum, the enduring virtual-energy sea that pulsates, producing periodic explosions that give rise to local universes. Universe-creating explosions (recurring "Bangs") are instabilities in the Metaverse's vacuum. The Bangs create pairs of particles and antiparticles, and the surviving surplus of particles populates the newborn universe's space-time. In time, gravitation pulls together these particles in galactic structures, and the kind of evolution we observe in our universe gets under way. It unfolds time after time.

The evolution of universes leads ultimately to quasars and black holes. Galaxies collapse on themselves as black holes form at their center, such as the black hole at the center of our Milky Way galaxy that was discovered recently. Sooner or later all galaxies "evaporate" in super-galactic black holes. These lead to further explosions—"star-bursts" of this kind have been observed—and these may prove to be the Bangs of subsequent universes.

Notwithstanding technical disagreements among different cosmological scenarios, most cosmologists agree that we live in a multiverse rather than a universe. Local universes evolve, die back, and coexist with, or are succeeded by, other universes in the embrace of a vast, temporally (if not necessarily spatially) infinite Metaverse. If these universes had no causal contact with one another, each of them would start with an accidental configuration of its basic laws and constants, and such a configuration, we have seen, has negligible chances of giving rise to complex systems such as living organisms. If we were to assume that at its birth our universe was not in causal contact with precursor universes, we would not be able to find natural causes for its astonishing propensity to bring forth life. Scientists could only marvel at the incredible serendipity that life could arise and evolve on Earth, and possibly elsewhere in this universe.

Instead of marveling at this improbable scenario, we can now explore the possibility that at its birth our universe *was* informed by a precursor universe. According to this cosmological conception, all universes leave their traces in the vacuum that embeds them, much as ships leave their traces in the sea on which they sail. These holographic traces do not cancel out as new universes are born; they superpose and accumulate. In consequence, there is an ongoing transfer of information between local universes: the "Bangs" of later universes are informed by the traces of their precursors. As the parameters of later universes become tuned to the processes that unfolded in the earlier universes, later universes neither collapse back on themselves shortly after their birth nor expand so fast that only a dilute gas of particles survives. They evolve more and more efficiently, and hence further and further than their predecessors.

As we have seen, our universe has laws and constants that are highly tuned to the evolution of life, indicating that when it was born, the primordial fluctuations of the vacuum are unlikely to have been random. They were precise, and this suggests that they were not accidental. So the logical conclusion regarding the origins of *our* universe is that the vacuum in which it arose was modulated by the traces of a prior universe.

What about the origins of the universe that preceded ours, and of all universes before that? *How did the Metaverse itself come into being?*

In considering this question, we should start with an important fact about complex systems: the fact that they are highly "initial-condition dependent." This means that their development is strongly influenced by the circumstances under which that development has been initiated. Our universe is a complex system; in fact, as far as we can tell, it is the most complex system there is. Its development must have been critically influenced by the conditions under which it was initiated—that is, by the fluctuating vacuum pre-space that exploded and created our universe's micro- and macro-structures, its particles and its galaxies.

We now apply the concept of initial-condition dependence to the Metaverse itself. The Metaverse's development also must have been critically influenced by its initial conditions. But prior universes could not have set these conditions, for the Metaverse was there before all universes. How, then, were the initial conditions of the Metaverse determined—*By what* . . . or is the question *By whom?* This is the deepest and greatest mystery of all—the mystery of the origins of the universe-generating process itself.

This greatest of all mysteries is "transempirical"; it is not amenable to resolution by reasoning based on observation and experiment. Yet one thing is clear: If it is unlikely that our fine-tuned universe would have originated in a series of random fluctuations, the mother universe that gave rise to a series of progressively evolving local universes is even more unlikely to have originated in that way. The Metaverse's pre-space was not only such that one universe could arise in it, but also such that an entire series of universes could. This could hardly have been a lucky fluke. We must admit that there must have been an original creative act, an act of "metaversal Design."

DESIGN OR EVOLUTION?

THE CREATIONIST CONTROVERSY IN A NEW LIGHT

The persistent debate among conservative Christians, Muslims, and Jews (the "creationists") and scientists and the science-minded public (the "evolutionists") centers on biological evolution. But on a deeper look, it concerns the universe itself in which life evolved—or in which it was created.

At first glance, the science community—and anyone believing that science discloses some basic truth about the nature of reality—is compelled to reject the hypothesis that living species are the way they are because they were designed to be that way . . . that they are the result of special acts of creation. Yet it is also evident that it is highly unlikely that living species could have come about through processes of random mutation and natural selection. Affirming this theory, the creationists claim, makes the entire doctrine of evolution misguided.

Mainline Darwinists expose themselves to the objection of the creationists by contending that random processes of evolution are adequate to explain the facts. Richard Dawkins, for example, claims that the living world is the result of processes of piecemeal trial and error without deeper meaning and significance. Like Weinberg, Dawkins claims that there is no purpose and meaning to this world. Therefore, there is no need to assume that it was purposefully designed.

Take cheetahs, he said. They give every indication of being superbly designed to kill antelopes. The teeth, claws, eyes, nose, leg muscles, backbone, and brain of a cheetah are all precisely what we should expect if God's purpose in creating cheetahs was to maximize deaths among antelopes. At the same time, antelopes are fast, agile, and watchful, apparently designed so they can escape cheetahs. Yet neither the one nor the other feature implies creation by special design: Dawkins tells us that this is just the way nature is. Cheetahs have a "utility function" to kill antelopes, and antelopes, to escape cheetahs. Nature itself is indifferent to their fate. Ours is a world of blind physical forces and genetic replication where some get hurt and others flourish. It has precisely the properties we would expect it to have if at bottom there was no design, no purpose, and no evil and no good, only blind and pitiless indifference.

Evidently, if this were the case, it would be hard to believe in an intelligent Creator. The God that created the world would have to be an indifferent God, if not actually a sadist who enjoys blood sports. It is more reasonable, according to Dawkins, to hold that the world just is, without reason and purpose. The way it is results from random processes played out within limits set by fundamental physical laws. The idea of design is superfluous. In this regard, Darwinists echo the French mathematician Pierre Laplace, who is reputed to have told Napoleon that God is a hypothesis for which there is no longer any need.

Creationists point out, however, that it is entirely improbable that all we see in this world, ourselves included, should be the result of random processes governed by impersonal laws. The tenet that everything evolved by blind chance out of common and simple origins is mere theory, they say, unsubstantiated by solid evidence. Scientists cannot come up with manifest proof for this theory of evolution: "You can't go into the laboratory or the field and make the first fish," said Tom Willis, director of the Creation Science Association for Mid-America. The world around us is far more than a chance concatenation of disjoined elements; it exhibits meaning and purpose. This implies design.

The creationist position would be the logical choice if cutting-edge evolutionary theory asserted that the origin of living species was truly the product of blind chance. But it does not. As we have seen, post-Darwinian biology has discovered that biological evolution is not merely the outcome of chance mutations exposed to natural selection. The coevolution of all things with all other things in the planet's web of life is a systemic process with a built-in dynamic. It is part of the evolution of the universe from particles to galaxies and stars with planets. On Earth this evolution produced physical, chemical, and thermal conditions that were just right for the stupendous processes of biological evolution to take off. Such conditions could have come about only in a universe governed by precisely coordinated laws and regularities. Even a minute difference in these laws and constants would have foreclosed the emergence of life forever.

Thus, the debate between creationists and evolutionists shifts from the question regarding the origins of life to the question concerning the origins of the universe. In the last analysis, it shifts to the origins of the Metaverse in which our universe arose. *Could it be that the Metaverse, the mother of our universe and of all universes past, present, and future, has been purposefully designed so that it could produce universes that give rise to life?* For creationists, this is the simplest and most logical assumption. Evolutionists cannot object: evolution, being an irreversible process, must have had a beginning, and that beginning must be accounted for. It could not have been something out of nothing—a "free lunch"!

In the final count, the evolutionist/creationist controversy has no point. The question "Design or evolution?" poses a false alternative. Design and evolution do not exclude each other; indeed, they require one another. The Metaverse is unlikely to have come into existence out of nothing, as a result of pure chance. And if it was designed, it was evidently so designed that it could evolve. The truth of the matter is not "design *or* evolution." It is "design *for* evolution."

Where is the universe going? We now reverse the direction of our inquiry. Instead of moving back in time, we move forward. In a coherently evolving universe this, too, is possible. The question we ask is: Where is the evolution of this universe, and of all universes in the Metaverse, leading—to what ultimate state or condition?

In contemplating this question we must realize that we are querying destiny and not fate. There is a fundamental difference between a point of origin and a point of destiny. A point of origin is in the past, and must be assumed to have been a definite and unique state. A point of destiny will be likewise a definite and unique state when it is reached—but it will not be that *until* it is reached. Much like the multipotentiality of the quantum that is free to choose its state until an interaction collapses its wave function, the cosmos will not have a determinate final state until it actually reaches that state. Not being classically mechanistic, it is undetermined as regards the choice of its ultimate state. The cosmos has various possibilities for its evolution.

The past is a stubborn fact, established once and for all, but the future is not. It is open, even if it is not *entirely* open. Ours is, after all, not a random, haphazard world, but one that evolves according to coherent laws and constants. This evolution is both self-consistent and irreversible. Its processes drive toward a definite *kind* of terminal state, but they do not predetermine one unique state as the only possible outcome.

Processes that head toward a final state that was not determined in the beginning are known to systems theorists: they are processes governed by so-called strange or chaotic attractors. These attractors introduce an element of indeterminacy into the systems. Computer simulations show that processes governed by such attractors reach an end state that is likely to be different in detail every time the simulation is run.

A GAME THAT GENERATES ITS OWN GOAL

There is a simpler way than computer simulation to experience processes that lead to goal states that were not given at the beginning. It can be done by playing the particular variant of the popular parlor game twenty questions that was suggested by the physicist John Wheeler (though he had an abstruse problem of quantum physics in mind). In the usual version of this game, a person leaves the room and the others decide on a thing or object that the person is to guess. The latter can ask a maximum of twenty questions, and only "yes" or "no" answers can be given to each question. But each question narrows the scope of possibilities because it excludes alternative possibilities. For example, if the first question is "Is it living?" (as opposed to nonliving), a yes answer excludes all things other than plants, animals, insects, and simple organisms.

In the alternative version, a person leaves the room and the others, without telling him, agree *not* to agree on a given thing or object but pretend that they did. They must give consistent answers, however. Consequently, when the innocent interlocutor returns and asks, "Is it living?" and if the answer he or she gets is yes, then all subsequent answers must pretend that the thing to be guessed is a plant, an animal, or perhaps a microorganism. A skilled player can narrow the scope of possibilities in such a way that within twenty questions he or she identifies one definite answer—for example, the kitten next door. Yet that was not the goal when the game was started. There was no goal—the one that emerged was generated by the game itself!

Our universe evolves with a great deal of coherence and consistency; one thing entails another. When one choice is made, the cascade of consequences continues until the final state is reached. The choices themselves are not random; they are constrained by the laws and constants of the universe system. The evolution of the universe has no fixed goal, but it does have a definite direction: toward growing structure and complexity. The evolutionary process adds part to part in coherent and self-consistent wholes. These in turn become part of still other, more encompassing self-consistent wholes.

Being so fine-tuned for the evolution of complexity, our universe could not have been the first universe to arise in the Metaverse. And if it was not the first universe, it is also not likely to be the last. Other universes will come about in time. *What universes?* We can also shed light on this far-reaching—but no longer "far-out"—query.

The evolution of the Metaverse is cyclic but not repetitive. One universe informs another; there is progress from universe to universe. Thus, each universe is more evolved than the one before. The mother universe itself evolves from a random initial universe, to universes where the physical parameters are more and more tuned to the evolution of complexity. Cosmic evolution is toward universes where complex structures emerge, including structures that harbor evolved forms of life—and the evolved forms of mind that are presumably associated with all evolved forms of life.

The Metaverse evolves from local universes that are purely *physical* to universes that include life. These are physical-*biological* universes. And given that forms of mind are associated with forms of life, the cycle of universes leads from physical to physical-biological to physical-biological-*psychological* worlds.

Is reaching a physical-biological-psychological universe the deeper meaning of the evolution, and perhaps the very existence, of the Metaverse? Possibly, and perhaps even probably. But we cannot be certain. A definitive answer is foreclosed to science, and to any reasoning this side of mystical intuition and prophetic insight.

LIFE ON EARTH AND IN THE UNIVERSE

Is there life elsewhere in the universe? We move now to the next set of "great questions": questions that are still "great" but somewhat more modest. They are questions about the origins and destiny of life on Earth and in the cosmos. The first query concerns the prevalence of life. *Is life unique to this planet or does it exist throughout the universe?*

We have every reason to believe that the kind of life we know on Earth is not limited to this planet. Life arose here over four billion years ago, and since then it has been evolving inexorably, if highly discontinuously, building structure upon structure, system within and with system. We have no reason to doubt that wherever suitable conditions are present, processes of physical, physical-chemical, and ultimately biological and ecological self-organization are getting under way. And we have every reason to believe that suitable conditions have been and are present in many places. Astronomical spectral analysis reveals a remarkable uniformity in the composition of matter in stars and hence in the planets that are associated with stars. The most abundant elements are, in order of rank: hydrogen, helium, oxygen, nitrogen, and carbon. Of these, hydrogen, oxygen, nitrogen, and carbon are fundamental constituents of life. Where these occur in the right distribution and energy is available to start chains of reaction, complex compounds result. On many planets the active star with which the planet is associated furnishes such energy. The energy is in the form of ultraviolet light, together with electric discharges, ionizing radiation, and heat.

About four billion years ago, photochemical reactions took place in the upper regions of the young Earth's atmosphere, and the reaction products were transferred by convection to the surface of the planet. Electric discharges close to the surface deposited the products in the primeval oceans, where volcanic hot springs supplied further energy. The combination of energy from the Sun with energy stored below the surface catalyzed a series of reactions of which the end products were organic compounds. The same system-building process is no doubt able to unfold with local variations on other planets. Numerous experiments pioneered by the paleobiologist Cyril Ponnamperuma and others show that when conditions similar to those that were present on the primeval Earth are simulated in the laboratory, the very compounds emerge that form the basis of Earthly life.

There must be other planets with conditions similar to those on Earth. There are more than 10^{20} stars in our universe, and during their active phase they all generate energy. When these energies reach the planets associated with the stars, they are capable of fueling the photochemical reactions required for life. Of course, not all stars are in the active phase, and not all have planets with the right chemical composition, of the right size, and at the right distance.

Just how many potentially life-bearing planets are there? The estimates vary. Taking a conservative tack, the Harvard astronomer Harlow Shapley assumed that only one star in a thousand has planets and that only one of a thousand of these stars has a planet at the right distance from it (in our solar system, there are two such planets). He further supposed that only one out of a thousand planets at the right distance is large enough to hold an atmosphere (in our system, seven planets are large enough), and that only one in a thousand planets at the right distance and of the right size has the right chemical composition to support life. Even then there should be at least 100 million planets capable of supporting life in the cosmos.

The astronomer Su-Shu Huang made less limiting assumptions and reached an even more optimistic estimate. He took the time scales of stellar and biological evolution, the habitable zones of planets and related dynamic factors, and came to the conclusion that no less than five percent of all solar systems in the universe should be able to support life. This means not 100 million, but 100 *billion* life-bearing planets. Harrison Brown came up with a bigger number still. He investigated the possibility that many planetlike objects that are not visible exist in the neighborhood of visible stars—perhaps as many as sixty such objects more massive than Mars. In that case, almost every visible star possesses a partially or wholly invisible planetary system. Brown estimated that there are at least 100 billion planetary systems in our own galaxy alone—and there are 100 billion galaxies in this universe. If he is right, life in the cosmos is immensely more prevalent than has been previously estimated.

This optimistic estimate has been underscored by a finding of the

Hubble Space Telescope in December of 2003. The Space Telescope succeeded in measuring a highly controversial object in an ancient part of our galaxy. It was not known whether this object is a planet or a brown dwarf. It has turned out to be a planet, having two and a half times the mass of Jupiter. It has an estimated age of thirteen billion years, which means that it must have formed barely a billion years after the birth of our universe!

Planets keep forming with remarkable speed and abundance to this day. In May of 2004, astronomers trained the new Spitzer Space Telescope at a "star nursery" region of the universe known as RCW 49, and in one image uncovered three hundred newborn stars, some not more than one million years old. A closer look at two of the stars showed that they have faint planet-forming disks of dust and gas around them. The astronomers estimated that all 300 might harbor such disks. This is a surprising discovery. If planets form around many stars, and if they form so soon, they must be far more abundant than was previously estimated.

If life potentially exists in so many places in the universe, wouldn't intelligent life and even technological civilization also exist? The probabilities in this regard were first calculated by Frank Drake in 1960. The famous Drake equation gives the statistical probabilities of the existence in our galaxy of stars with planets; of planets with environments capable of sustaining life; of life on some of the life-friendly planets; of intelligent life on some of the actually life-bearing planets; and of advanced technological civilization produced by the intelligent life that evolved on these planets. Drake found that, given the large number of stars in our galaxy, as many as ten thousand advanced technological civilizations are likely to exist in the Milky Way galaxy alone.

The Drake equation was updated and elaborated by Carl Sagan and colleagues in 1979. Their computations claim that not ten thousand, but up to one million intelligent civilizations could exist in our galaxy. In the late 1990s, Robert Taormina applied these equations to a region within one hundred light-years from Earth and found that more than eight such civilizations should be present within "hailing distance" from us. And in light of the fact that planets started to form a billion years or so after the birth of the universe, these estimates must be revised upward once again.

Should we be hearing from an advanced extraterrestrial civilization before long? The chances of interplanetary communication are real. In the last fifteen years, twelve hundred Sun-like stars in our vicinity have been scrutinized by astronomers with ground-based telescopes, and their search has come up with ninety extrasolar planets. A particularly promising find was announced in June 2002: the planetary system known as 55 Cancri. It is within hailing distance: forty-one light-years from us. It appears to have a planet that resembles Jupiter in mass and in regard to orbit. Calculations indicate that 55 Cancri could also have rocky planets much like Mars, Venus, and Earth.

However, this is a relatively exceptional find. Most of the other solar systems in our neighborhood have alien planets in widely eccentric orbits, moving either too far from their host sun to sustain life or moving too close to it.

Even though planets appear to be highly abundant in this galaxy and elsewhere in the cosmos, planets capable of sustaining advanced forms of life could be relatively rare. According to Peter Ward and Donald Brownlee, radiation and heat levels are so high on most planets that the only forms of life that are likely to exist are a variety of bacteria deep in the soil. The odds against advanced technological civilization beyond Earth, they say, are astronomical. But even if planets with the right composition, the right distance from the host star, and the right orbit were rare in the universe, the existence of advanced civilizations could not be excluded. There are an astronomical number of stars and planets, so even if the odds are astronomically against such civilizations, they do not foreclose their actual existence, but only indicate that they would be less frequent.

Although the evolution of cellular and then multicellular organisms on suitable planets may take millions if not billions of years, life has most probably evolved to higher forms on some other planets, even if not on very many others. Under particularly favorable conditions, evolution is likely to lead to advanced organisms with an evolved brain and nervous system. And these organisms are likely to have an evolved consciousness capable of establishing advanced civilizations. This means that even if they are relatively rare, extraterrestrial civilizations are likely to exist, created by complex organisms on life-bearing planets.

In the informed universe, the existence of life, and also of advanced civilizations, is far more probable than in a noninformed universe. Through the A-field, life in any one place informs and facilitates the evolution of life in other places, so the evolution of life never starts from scratch. It is not at the mercy of lucky flukes of random mutations coming up with organisms that prove viable in changing environments.

The evolution of life on Earth did not rely on chance mutations, nor did it require the physical importation of organisms or proto-organisms from elsewhere in the solar system, as the "biological seeding" theories of the origins of life suggest. Instead, the chemical soup out of which the first proto-organisms arose was informed by the A-field-conveyed traces of extraterrestrial life. Life on Earth was not biologically, but rather *informationally* seeded.

A-field-conveyed interplanetary information is a subtle prompt that speeds up the evolution of complex systems. It favors the rise of advanced life-forms under suitable thermal and chemical conditions. Such information increases the chances that organisms evolve that are capable of creating a form of civilization.

Can the human brain pick up interplanetary information? Very likely it can, even though our everyday logic suppresses it for being strange, without evident sensory origin. Traces of it can emerge nonetheless in altered states of consciousness, where the censorship that filters incoming information is temporarily lifted.

At this crucial juncture in the evolution of human civilization, it would be of particular importance to open our minds to interplanetary information. Numerous civilizations are likely to exist in this galaxy, and in the 100 billion other galaxies of our universe. These civilizations must also have faced at some point the challenge of finding a way to live on their home planet without allowing their technologies to damage the natural cycles that make up their biosphere. If they survived, they met this challenge. But how did they achieve a condition of sustainability? The answer must be in the A-field. Getting an inkling of it may make the crucial difference between bumbling along in a fateful gamble with trial and error and moving with intuitive assurance toward solutions that have been already tried and tested—even if not here, but elsewhere in the universe.

THE FUTURE OF LIFE IN THE COSMOS

The reasonable certainty that life, even advanced forms of life, exists not just on Earth does not tell us that life will exist forever, whether on this or on other planets. The fact is that life cannot exist indefinitely in the cosmos: the physical resources required for carbon-based life—the only kind we know of—do not last forever.

The evolution of the known forms of life depends on a strictly limited range of temperatures and the presence of a specific variety of chemical compounds. These factors, as we have seen, are likely to exist on a number of planets in this and other galaxies, on planets that have the right chemical and thermal conditions, situated at the right distance from their active star. But whether such planets are highly abundant or relatively rare, the conditions they provide for the sustenance of life are limited in time. The principal reason is that the active phase of the stars whose radiation drives the processes of life does not last forever. Sooner or later stars exhaust their nuclear fuel, and then they either shrink to the white dwarf stage or fly apart in a supernova explosion. The population of active stars is not infinitely replenished in this universe. Even if new stars keep forming from interstellar dust, a time must come when no further stars are born.

Even if the time spans are mind-boggling, the limitations are real. About 10^{12} (one trillion) years from now, all the stars that remain in our universe will first have converted their hydrogen into helium—the main fuel of the supercompacted but still luminous white dwarf state—and then will have exhausted their supply of helium. We have already been able to observe that the galaxies constituted of such stars take on a reddish tint, then—when their stars cool still further—fade from sight altogether. As energy is lost in the galaxies through gravitational radiation, individual stars move closer together. The chance of collision among them increases, and the collisions that occur precipitate some stars toward the center of their galaxies and expel others into extragalactic space. As a result, the galaxies diminish in size. Galactic clusters also shrink, and in time both galaxies and galactic clusters implode into black holes. At the time horizon of 10^{34} years, all matter in our universe will be reduced to radiation, positronium (pairs of positrons and electrons), and compacted nuclei in black holes.

Black holes themselves decay and disappear in a process Stephen Hawking calls evaporation. A black hole resulting from the collapse of a galaxy evaporates in 10^{99} years, while a giant black hole containing the mass of a galactic supercluster vanishes in 10^{117} years. (If protons do not decay, this span of time expands to 10^{122} years.) Beyond this humanly inconceivable time horizon, the cosmos contains matter particles only in the form of positronium, neutrinos, and gamma-ray photons.

Whether the universe is expanding (open), expanding and then contracting (closed), or balanced in a steady state, the complex structures required for the known forms of life vanish before matter itself supercrunches, or evaporates.

In the late phases of a *closed universe*—one that ultimately collapses back on itself—the universe's background radiation increases gradually but inexorably, subjecting living organisms to mounting temperatures. The wavelength of radiation contracts from the microwave region into the region of radio waves, and then into the infrared spectrum. When it reaches the visible spectrum, space is lit with an intense light. At that time all life-bearing planets are vaporized, along with every other object in the vicinity.

In an *open universe* that expands indefinitely, life dies out because of cold rather than heat. As galaxies continue to move outward, many active stars complete their natural life cycle before gravitational forces bunch them close enough to create a serious risk of collision. But this does not improve the prospects of life. Sooner or later all the active stars of the universe exhaust their nuclear fuel and then their energy output diminishes. The dying stars either expand to the red giant stage, swallowing up their inner planets, or settle into lower luminosity levels on the way to becoming white dwarfs or neutron stars. At these diminished energy levels, they are too cold to sustain whatever organic life may have evolved on their planets.

A similar scenario holds in a *steady-state universe*. As active stars approach the end of their life cycle, their energy output falls below the threshold where life can be supported. Ultimately a lukewarm, evenly distributed radiation fills space, in a universe where the remnants of matter are random occurrences. This universe is incapable of maintaining the flame of a candle, not to mention the complex irreversible reactions that are the basis of life.

Whether our universe expands and then contracts, expands infinitely, or reaches a steady state, the later stages of its evolution will wipe out the known forms of life.

This is a dismal picture, but it is not the whole picture. The whole picture is not limited to our own finite universe; there is also a temporally (whether or not also spatially) infinite or quasi-infinite Metaverse. And life in the Metaverse need not end with the devolution of local universes. While life in each local universe must end, it can evolve again in the universes that follow.

If evolution in each local universe starts with a clean slate, the evolution of life in local universes is a Sisyphean effort: it breaks down and starts again from scratch, time after time. But local universes are not subject to this ordeal. In each universe, complex systems leave their traces in the vacuum, and the informed vacuum of one universe informs the evolution of the next. Consequently, each universe creates conditions favorable to the evolution of life in successive universes. In each successive universe, life evolves more and more efficiently, and thus in equal times evolves further and further.

This is a cyclical process with a learning curve. Each universe starts without life, evolves life when some planets become capable of supporting it, and wipes it out when planetary conditions pass beyond the life-supporting stage. But the vacuum shared by all the universes records and conserves the wave-form traces of the life that evolved in each universe! The vacuum becomes more and more informed *with* life, and therefore more and more informing *of* life.

Cyclically progressive evolution in the Metaverse offers a positive prospect for the future of life: it can continue in one universe after another. And it can evolve further, in universe after universe.

What can we say about the super-evolved forms of life that would come about in the mature stages of mature universes? Since the course of evolution is never precisely predictable, we can actually say very little. All we can surmise is that mature organisms in mature universes will be more complex, more coherent, and more whole than the forms of life familiar to us. In most other respects they could be as different from the organisms we know on Earth as humans are different from the protozoan slime that once populated the primeval seas of this planet.

A Footnote on Reality

We end the first part of our explorations of the informed universe with a question that is meaningful but decidedly not modest: a question about the nature of reality. We have already seen how our universe and possibly myriad other universes in the Metaverse came into being, how they evolve and devolve, and how they periodically give rise to the complex systems we call living. What do these stupendous processes tell us about the fundamental nature of reality? What is it about this universe that is primary and what is merely secondary, arising out of the reality of the primary?

The answer to this age-old question is now relatively straightforward. *The primary reality is the quantum vacuum, the energy- and information-filled plenum that underlies our universe, and all universes in the Metaverse.*

This answer corresponds to an ancient insight: that the universe we observe and inhabit is a secondary product of the energy sea that was there before there was anything there at all. Hindu and Chinese cosmologies have always maintained that the things and beings that exist in the world are a concretization or distillation of the basic energy of the cosmos, descending from its original source. The physical world is a reflection of energy vibrations from more subtle worlds that, in turn, are reflections of still more subtle energy fields. Creation, and all subsequent existence, is a progression downward and outward from the primordial source.

In Indian philosophy the ultimate end of the physical world is a return to Akasha, its original subtle-energy womb. At the end of time as we know it, the almost infinitely varied things and forms of the manifest world dissolve into formlessness, living beings exist in a state of pure potentiality, and dynamic functions condense into static stillness. In Akasha, all attributes of the manifest world merge into a state that is beyond attributes: the state of Brahman.

Although it is undifferentiated, Brahman is dynamic and creative. From its ultimate "being" comes the temporary "becoming" of the manifest world, with its attributes, functions, and relationships. The cycles of *samsara*—of being-to-becoming and again of becoming-to-being—are the *lila* of Brahman: its play of ceaseless creation and dissolution. In Indian philosophy, absolute reality is the reality of Brahman. The manifest world enjoys but a derived, secondary reality and mistaking it for the real is the illusion of *maya*. The absolute reality of Brahman and the derived reality of the manifest world constitute a co-created and constantly co-creating whole: this is the *advaitavada* (the nonduality) of the universe.

The traditional Eastern conception differs from the view held by most people in the West. In the modern commonsense conception, reality is material. The things that truly exist are bits or particles of matter. They can form into atoms, which can further form into molecules, cells, and organisms—as well as into planets, stars, stellar systems, and galaxies. Matter moves about in space, acted on by energy. Energy also enjoys reality (since it acts on matter), but space does not: space is merely the backdrop or the container against which, or in which, material things trace their careers.

This typically Western view is a heritage of the Newtonian world-concept. According to Newton, space is a mere receptacle and it is passive in itself; it conditions how things actually behave but does not act on them directly. Although it is empty and passive, space is nonetheless real: it is an objective element in the universe. Subsequently, a number of philosophers, including Gottfried Leibniz and Immanuel Kant, contested the reality Newton gave to space. In these views, space is nothing in itself; it is merely the way we order relationships among real things. Space itself is not experienced, said Kant; it is only the precondition of experience.

The view that space is empty and passive, and not even real to boot, is in complete opposition to the view we get from contemporary physics. Even if physicists typically refuse to speculate on the ultimate nature of reality (many hold such questions beyond the scope of their discipline), it is clear that what they describe as the unified vacuum—the seat of all the fields and forces of the physical world—is in fact the primary reality of the universe. Out of it have sprung the particles that make up our universe, and when the last of the supergalactic black holes "evaporates," it is into it that the particles fall back again. What we think of as matter is but the quantized, semi-stable bundling of the energies that spring from the vacuum. In the last count matter is but a waveform disturbance in the nearly infinite energy-sea that is the fundamental medium—and hence the primary reality—of this universe, and of all universes that ever existed and will ever exist.

EIGHT

Consciousness:
Human and Cosmic

We continue now to query the informed universe. If this universe is the cornerstone of an integral theory of everything, it should provide answers to a further set of questions, centered not on the manifest facts of nature and life, but on the more subtle facts of consciousness. The questions we ask here are about:

- —the roots of the phenomenon we know as consciousness
- —the wider range of the information that reaches and forms our (and any other) consciousness
- —the next evolution of human consciousness
- —the likelihood that consciousness exists elsewhere in the universe
- —the possibility that our consciousness is immortal.

THE ROOTS OF CONSCIOUSNESS

Contrary to a widespread belief, consciousness is not a uniquely human phenomenon. Although we know only human consciousness (indeed, by direct and indubitable experience we know only our *own* consciousness), we have no reason to believe that consciousness would be limited to me and to you and to other humans.

The kind of evidence that could demonstrate the limitation of consciousness to humans regards the brain: it would be evidence that the human brain has specific features by virtue of which it *produces* consciousness. Notwithstanding the view advanced by materialist scientists and philosophers that the physical brain is the source of consciousness, there is no evidence of this kind. Clinical and experimental evidence speaks only to the fact that brain function and state of consciousness are correlated, so when brain function ceases, consciousness (usually) ceases as well. We should specify "usually," since there are exceptions to this: in some well-documented cases—among others, those of patients suffering cardiac arrest in hospitals—individuals have had detailed and subsequently clearly recalled experiences during the time their EEG showed a complete absence of brain function.

Functional MRI (magnetic resonance imaging) and other techniques show that when particular thought processes occur, they are associated with metabolic changes in specific areas of the brain. They do *not* show, however, how the cells of the brain that produce proteins and electrical signals could also produce sensations, thoughts, emotions, images, and other elements of the conscious mind . . . how the brain's network of neurons would produce the qualitative sensations that make up our consciousness.

The fact that a high level of consciousness, with articulated images, thoughts, feelings, and rich subconscious elements, is *associated* with complex neural structures is not a guarantee that such consciousness is *due* to these structures. In other words the observation that brain function is *correlated* with consciousness does not ensure that the brain *creates* consciousness.

ALTERNATIVE APPROACHES TO THE BRAIN-MIND PROBLEM

The view that consciousness is produced in and by the brain is just one of the many ways philosophically inclined people have envisaged the relationship between the physical brain and the conscious mind. It is the *materialist* way. It maintains that consciousness is a kind of by-product of the survival functions the brain performs for the organism. As organisms become more complex, they require a more complex "computer" to steer them so they can get the food, the mate, and the related resources they need in order to survive and reproduce. At a given point in this development, consciousness appears. Synchronized neural firings and transmissions of energy and chemical substances between synapses produce the qualitative stream of experience that makes up our consciousness. Consciousness is not primary in the world; it is an "epi-phenomenon" generated by a complex material system: the human brain.

The materialist way of envisaging the relationship of brain and mind is not the only way. Philosophers have also outlined the *idealist* way. In the idealist perspective, consciousness is the first and only reality; matter is but an illusion created by our mind. This assumption, while outlandish on first sight, makes eminent sense as well: after all, we do not experience the world directly; we experience it only through our consciousness. We normally assume that there is a qualitatively different physical world beyond our consciousness, but that may be an illusion. Everything we experience could be part of our consciousness. The material world could be merely our creation as we try to make sense of the flow of sensations in our consciousness.

Then there is the *dualist* way of conceiving of the relationship between brain and consciousness, matter and mind. According to dualist thinkers, mind and matter are both fundamental, but they are entirely different, not reducible one to the other. The manifestations of consciousness cannot be explained by the organism that manifests them, not even by the staggeringly complex processes of the human brain—the brain is only the seat of consciousness and is not identical with it.

In the history of philosophy, materialism, idealism, and dualism were the principal ways of conceiving the relationship between brain and mind. Materialism is still dominant today. Adherence to it poses vexing problems. As the consciousness philosopher David Chalmers put it, the problem it faces is how "something as immaterial as consciousness" can arise from "something as unconscious as matter." In other words, *how can matter generate mind*? How the brain operates is a comparatively "soft" problem that neurophysiologists will no doubt solve step by step. But the question regarding the way in which "immaterial consciousness" arises out of "unconscious matter" cannot be answered by brain research, for that deals only with "matter," and matter is not conscious. This is the "hard" problem.

Consciousness researchers of the materialist school admit to being greatly perplexed by it. The philosopher Jerry Fodor points out that "nobody has the slightest idea how anything material could be conscious. Nobody even knows what it would be like to have the slightest idea about how anything could be conscious." But philosophers who do not take the materialist stance are not as disturbed. Peter Russell, for example, says that Chalmers's problem is not just hard; it's impossible. Fortunately, Russell adds (and we can agree), it does not need to be solved, for it is not a real problem. We do not need to explain how unconscious matter generates immaterial consciousness, because matter is not entirely unconscious, nor is consciousness fully divorced from matter.

We know that the "stuff" of the neurons in the brain comprises quanta in complex configurations. But quanta are not mere unconscious matter! They stem from the basic constituents of the complex fields that underlie the cosmos, and they are not devoid of the qualities we associate with consciousness. As leading physicists such as Freeman Dyson and philosophers of the stature of Alfred North Whitehead have pointed out, even elementary particles are endowed with a form and level of (proto) consciousness. *To some extent and in some ways, all matter is conscious, and no consciousness is categorically immaterial.* And if so, there is no categorical divide between matter and mind.

David Chalmers's "hard" problem evaporates. Conscious matter at a lower level of organization (the neurons in the brain) generates conscious matter at a higher level of organization (the brain as a whole). This does away with the hard problem of the materialist view without doing the kind of violence to our everyday apprehension of the world that idealism does (according to which all is mind, and nothing but mind). It also does away with the problem of dualism—one that is just a shade less "hard" than that of materialism—because if matter and mind interact (as they must interact in the brain), then we must still say how "something as unconscious as matter" can act on, and be acted on by, "something as immaterial as consciousness."

The "ism" by which we can best identify the emerging solution to the classical brain/mind problem is *evolutionary panpsychism*. Panpsychism is the philosophical position that claims that all of reality has a mental aspect: psyche is a universal presence in the world. Qualifying "panpsychism" with "evolutionary" means that we do not claim that psyche is present throughout reality in the same way, at the same level of development. We say that psyche evolves, the same as matter. But we affirm that both matter and mind—*physis* and *psyche*— were present from the beginning: they are both fundamental aspects of reality.

In affirming that in the course of time *physis* and *psyche* evolved together, we do not reduce all of reality to structures made up of in-themselves inert and insentient material building blocks (as in materialism), nor do we assimilate all of reality to a qualitative nonmaterial mind (as in idealism). We take both matter and mind as fundamental elements of reality but (unlike in dualism) we do not claim that they are radically separate; we say that they are but different aspects of the *same* reality. What we call "matter" is the aspect we apprehend when we look at a person, a plant, or a molecule from the *outside;* "mind" is the readout we get when we look at the same thing from the *inside.*

Of course, for each of us the inside view is available only in regard to our own brain. It is not the complex network of neurons that we see when we inspect what we assume to be the felt contents of our brain, but a complex stream of ideas, feelings, intentions, and sensations. This is the stream of our consciousness with its manifold conscious and also subconscious elements. But it is not this stream that we apprehend when we inspect anybody else's brain-mind. What we get is the neuroscientist's view of a network of neurons firing in complex loops and sequences.

The limitation of the inside view to our own brain does not mean that we alone are conscious and everyone else is but a neurophysiological mechanism operating within a biochemical system. Both views—the outside as well as the inside—must be present in all human beings. And not only in all humans, but also in all other biological organisms. And not only in organisms, but also in all the systems that arise and evolve in nature, from atoms to molecules, to macromolecules, to ecologies. In the great chain of evolution, there is nowhere we can draw the line, nowhere we could say: below this there is no consciousness, and above there is.

This panpsychist concept has been espoused by philosophers over the ages, in modern times most eloquently by Alfred North Whitehead. It was also affirmed by the *Apollo* astronaut Edgar Mitchell. According to Mitchell, all things in the universe have a capacity to "know." Less evolved forms of matter, such as molecules, exhibit more rudimentary forms of knowing—they "know" to combine into cells. Cells "know" to reproduce and fight off harmful intruders; plants "know" to turn toward the sun, birds to fly south in winter. The higher forms of knowing, such as human awareness and intention, have their roots in the cosmos; they were there right from the start, at the birth of our universe.

The idea that mind and knowing are universal in nature is shared by Freeman Dyson. "Matter in quantum mechanics," he said, "is not an inert substance but an active agent, constantly making choices between alternative possibilities. . . . It appears that mind, as manifested by the capacity to make choices, is to some extent inherent in every electron."

In the final count we must recognize that all the things that arise and evolve in the universe have both a matter-aspect and a mind-aspect. All things in the world—quanta and galaxies, molecules, cells, and organisms—have "materiality" as well as "interiority." Matter and mind are not separate, distinct realities; they are aspects of a deeper reality that has both an external matter-aspect, and an internal mind-aspect.

THE WIDER INFORMATION OF CONSCIOUSNESS

Is the information reaching our consciousness limited to our bodily senses—do we see the world through "five slits in the tower"? Or can we "open the roof to the sky"? The informed universe gives us not only a new view of the world, but also a new view of life and of mind. It permits our brains and minds to access a broad band of information, well beyond the information conveyed by our eyes and ears. We are, or can be, literally "in touch" with almost any part of the world, whether here on Earth or beyond in the cosmos.

When we do not repress the corresponding intuitions, we can be informed by things as small as a particle or as large as a galaxy. This, we have seen, is the finding of psychiatrists and psychotherapists who place their patients in an altered state of consciousness and record the impressions that surface in their minds. It was also astronaut Mitchell's outer-space experience. In a higher state of consciousness, he remarked, we can enter into deep communication with the universe. In these states the awareness of every cell of the body coherently resonates with what Mitchell identifies as "the holographically embedded information in the quantum zero-point energy field."

We can reconstruct how this "broad-band" information reaches our mind. We have seen that, according to the new physics, the particles and atoms—and the molecules, cells, organisms, and galaxies—that arise and evolve in space and time emerge from the virtual energy sea that goes by the name of quantum vacuum. These things not only originate in the vacuum's energy sea; they continually interact with it. They are dynamic entities that read their traces into the vacuum's A-field, and through that field enter into interaction with each other. A-field traces—the holograms they create—are not evanescent. They persist and inform all things, most immediately the same kind of things that created them.

This holds true for our body and brain as well. All we experience in our lifetime—all our perceptions, feelings, and thought processes—have cerebral functions associated with them. These functions have wave-form equivalents, since our brain, like other things in space and time, creates information-carrying vortices—it "makes waves." The waves propagate in the vacuum and interfere with the waves created by the bodies and brains of other people, giving rise to complex holograms. Generations after generations of humans have left their holographic traces in the A-field. These individual holograms integrate in a super-hologram, which is the encompassing hologram of a tribe, community, or culture. The collective holograms interface and integrate in turn with the super-superhologram of all people. This is the collective information pool of humankind.

We can read the information carried by these holograms. On the principle of "like informs like," we can read first of all the information carried by the hologram of our own brain and body. Reading *out* what we have read *into* the field is the physical basis of long-term memory. It removes the limitation on information storage by a brain enclosed in a finite cranium. We can read out anything and everything that we have read into the field—we can literally "re-call" from it all the things we have ever experienced.

Not only we ourselves, but others also can read out at least some of what we have read into the A-field. This is because the hologram of our body and brain can "conjugate" with the holograms of other people, especially people who are related to us and with whom we have an emotional bond. Aside from cases of clairvoyance and mystical or prophetic insight, the readout is not in the form of explicit words or events, but rather in the form of intuitions and sensations. The most widespread and hence familiar among these are "twin pain" and the sudden revelatory intuitions of mothers and lovers when their loved ones are hurt or undergo a traumatic experience.

In the everyday context, of course, our readout is restricted to our own read-in. This restriction is fortunate: it is a precondition of conserving our sanity. If the experience of many people reached us simultaneously and frequently, we would be overwhelmed—we could not sort out the information. Given the holopattern selectivity of our brain's readout—the limited way our own hologram meshes with the hologram of others—we are not swamped by the enormity of the information in the A-field.

This does not mean that human experience must be limited to five slits in the tower. By entering an altered state of consciousness, we can open the roof to the sky, but we must be prepared to cope with the information that is then reaching us.

THE NEXT EVOLUTION OF HUMAN CONSCIOUSNESS

Our consciousness is not a permanent fixture: cultural anthropology testifies that it developed gradually in the course of millennia. In the thirty- or fifty-thousand-year history of modern man, the human body did not change significantly, but human consciousness did. It evolved from simpler beginnings and, if humankind survives long enough, it will evolve further.

Different levels of human consciousness, with progressive evolution from the lowest to the highest, were envisaged by almost all the great spiritual traditions. For example, some Native American cultures (the Mayan, Cherokee, Tayta, Xingue, Hopi, Inca, Seneca, Inuit, and Mapuche traditions) hold that we are presently living under the Fifth Sun of consciousness and are on the verge of the Sixth Sun. The Sixth Sun will bring a new consciousness and with it a fundamental transformation of our world.

A number of thinkers attempted to define the specific steps or stages in the evolution of human consciousness. The Indian sage Sri Aurobindo considered the emergence of superconsciousness in some individuals as the next step; in a similar vein the Swiss philosopher Jean Gebser spoke of the coming of four-dimensional integral consciousness, rising from the prior stages of archaic, magical, and mythical consciousness. The American mystic Richard Bucke portrayed cosmic consciousness as the next evolutionary stage of human consciousness, following the simple consciousness of animals and the self-consciousness of contemporary humans. Ken Wilber's six-level evolutionary process leads from physical consciousness pertaining to nonliving matter energy through biological consciousness associated with animals and mental consciousness characteristic of humans to subtle consciousness, which is archetypal, transindividual, and intuitive. It leads in turn to causal consciousness and, in the final step, to the ultimate consciousness called Consciousness as Such. And Chris Cowan's and Don Beck's colorful spiral dynamics sees contemporary consciousness evolving from the strategic "orange" stage that is materialistic, consumerist, and success-, image-, status-, and growth-oriented; to the consensual "green" stage of egalitarianism and orientation toward feelings, authenticity, sharing, caring, and community; heading toward the ecological "yellow" stage focused on natural systems, self-organization, multiple realities, and knowledge; and culminating in the holistic "turquoise" stage of collective individualism, cosmic spirituality, and Earth changes.

Ideas such as these differ in specific detail, but they have a common thrust. Consciousness evolution is from the ego-bound to the transpersonal form. If this is so, it is a source of great hope. Transpersonal consciousness is open to more of the information that reaches the brain than the dominant consciousness of today. This could have momentous consequences. It could produce greater empathy among people, and greater sensitivity to animals, plants, and the entire biosphere. It could create subtle contact with other parts of the cosmos. It could change our world.

A society hallmarked by transpersonal consciousness is not likely to be materialistic and self-centered; it would be more deeply and widely informed. Under the impact of a more evolved consciousness, the system of nation-states would transform into a more inclusive and coordinated system with due respect for diversity and the right of all peoples and cultures to self-determination. Economic systems would remain diversified but not fragmented; they would combine local autonomy with global coordination and pursue goals that serve all the peoples and countries of the world, whatever their creed, level of economic development, population size, and natural resource endowment. As a result, disparities in wealth and power would be moderated and frustration and resentment would diminish, together with crime, terrorism, war, and other forms of violence. Societies would become more peaceful and sustainable, offering a fair chance of life and well-being to all their members, living and yet to be born.

Will this condition, in today's perspective distinctly utopian, actually come about? This we cannot say: evolution is never fully predictable. All we can say is that if humankind does not destroy its life-supporting environment and decimate its populations, the dominant consciousness of a critical mass will evolve from the ego-bound to the transpersonal stage. This evolution is certain to leave its mark on people and societies. When our children and grandchildren graduate to transpersonal consciousness, an era of peace, fairness, and sustainability could dawn for humanity.

COSMIC CONSCIOUSNESS

We can now take another step in our exploration of the informed universe: a step *beyond* the consciousness associated with organisms and other complex systems. *Could the cosmos itself possess consciousness in some form?*

Through the ages, mystics and seers have affirmed that consciousness is fundamental in the universe. Seyyed Hossein Nasr, a medieval Islamic scholar and philosopher, wrote, "[T]he nature of reality is none other than consciousness. . . . " Sri Aurobindo concurred: "[A]ll is consciousness—at various levels of its own manifestation . . . this universe is a gradation of planes of consciousness." Scientists have occasionally joined the ranks of the mystics. Sir Arthur Eddington noted, "[T]he stuff of the universe is mind-stuff . . . the source and condition of physical reality." And the Nobel laureate biologist George Wald said that mind, rather than emerging as a late outgrowth in the evolution of life, has existed always.

Nearly two-and-a-half thousand years ago Plato recognized that in regard to ultimate questions there can be no certainty: the best we can do is to find the most likely story. In the contemporary context, the likeliest story is that consciousness is universal in nature. Its roots extend to the heart of physical reality: to the quantum vacuum. We know that this subtle virtual energy sea is the originating ground of the bound-energy wave-packets we view as matter, and we now have excellent grounds to assume that it is the originating ground of mind as well.

How could we tell that the vacuum is not only the seat of a super-dense virtual energy field from which spring wave-packets we call matter, but also a cosmically extended proto- or root-consciousness? There is no way we could tell by ordinary sensory experience. First, because we cannot observe vacuum fields, we can only conclude their existence by reasoning from the things we *can* observe. Second, because consciousness is "private," we cannot ordinarily observe it in anyone other than ourselves. The claim that the vacuum is both a virtual energy field and a field of proto-consciousness is condemned to remain hypothetical, even if supported by indirect evidence.

There are, however, positive approaches we can take. To begin with, even if we cannot directly observe consciousness in the vacuum, we could attempt an experiment. We could enter an altered state of consciousness and identify ourselves with the vacuum, the deepest and most fundamental level of reality. Assuming that we succeed (and psychotherapists tell us that in altered states people can identify with almost any part or aspects of the universe), would we experience a physical field of fluctuating energies? Or would we experience something like a cosmic field of consciousness? The latter is much more likely. We have already noted that when we experience anybody else's brain "from the outside," we do not experience his or her consciousness—at the most we experience gray matter consisting of complex sets of neurons firing in complex sequences. But we know that when we experience *our* brain "from the inside," we experience not neurons, but the qualitative features that make up our stream of consciousness: thoughts, images, volitions, colors, shapes, and sounds. Would not the same hold true when we project ourselves into a mystical union with the vacuum?

This is not just a fanciful supposition: there is indirect yet significant evidence for it. It comes from the farther reaches of contemporary consciousness research. Stanislav Grof found that in deeply altered states of consciousness, many people experience a kind of consciousness that appears to be that of the universe itself. This most remarkable of altered-state experiences surfaces in individuals who are committed to the quest of apprehending the ultimate grounds of existence. When the seekers come close to attaining their goal, their descriptions of what they regard as the supreme principle of existence are strikingly similar. They describe what they experience as an immense and unfathomable field of consciousness endowed with infinite intelligence and creative power. The field of cosmic consciousness they experience is a cosmic emptiness—a void. Yet, paradoxically, it is also an essential fullness. Although it does not feature anything in a concretely manifest form, it contains all of existence in potential. The vacuum they experience is a plenum: nothing is missing in it. It is the ultimate source of existence, the cradle of all being. It is pregnant with the possibility of everything there is. The phenomenal world is its creation: the realization and concretization of its inherent potential.

Basically, the same kind of experience is recounted by people who practice yoga and other forms of deep meditation. The Indian Vedic tradition, for example, regards consciousness not as an emergent property that comes into existence through material structures such as the brain and the nervous system, but as a vast field that constitutes the primary reality of the universe. In itself, this field is unbounded and undivided by objects and individual experiences, but it can be experienced in meditation when the gross layers of the mind are stripped away. Underlying the diversified and localized gross layers of ordinary consciousness there is a unified, nonlocalized, and subtle layer: "pure consciousness."

According to traditional cosmologies, the universe's undifferentiated, all-encompassing consciousness separates off from its primordial unity and becomes localized in particular structures of matter. In the new scientific context we can specify that the proto-consciousness of the quantum vacuum becomes localized and articulated as particles emerge from it and evolve into atoms and molecules. On life-bearing planets they evolve further into cells, organisms, and ecologies. The human mind, associated with the highly evolved human brain, is a high-level articulation of the cosmic consciousness that, emerging from the vacuum, infuses all things in space and time.

IMMORTALITY AND REINCARNATION

Last but not least we ask the most exciting of all the great questions people have ever asked. *Could our consciousness survive the physical demise of our body?*

We can also shed light on this perennial question, but not by applying the usual methods of the sciences. It does not help to examine the human brain, for if consciousness continues to exist when brain function ceases, it is no longer associated with the brain. It is more to the point to look at the evidence furnished by instances where consciousness is no longer directly linked with the brain. This is the case in near-death experiences, out-of-body experiences, past-life experiences, some varieties of mystical and religious experiences, and, perhaps most significant of all, the experiences of after-death communication. Until recently, scientists could not cope with such "paranormal" experiences; they did not fit into the materialist scheme of scientific thinking. But the informed universe is not the materialist's kind of universe. Let us take a fresh look at the phenomena, and see what kind of explanation we can now find for them.

Immortality

In near-death experiences, out-of-body experiences, past-life experiences, and various mystical and religious experiences, people perceive things that were not conveyed by their eye, ear, or other bodily senses. As we have seen, in NDEs the brain can be clinically dead, with the EEG "flat," and yet people can have clear and vivid experiences that, when they come back from the portals of death, they can recall in detail. In OBEs people can "see" things from a point in space that is removed from their brain and body, while in mystical and religious transport, experiencing subjects have the sense of entering into union with something or someone larger than themselves, and indeed larger or higher than the natural world. Although in some of these experiences the consciousness of individuals is detached from their physical brain, their experiences are vivid and realistic. Those who undergo them seldom doubt that they are real.

In addition to NDEs, OBEs, and mystical experiences, another remarkable form of experience has surfaced in recent years: experience in which there appears to be contact and communication with people who are no longer alive. This kind of experience became known as ADC: after-death communication.

Many people seem to experience after-death communication; the NDE researcher Raymond Moody collected a wide variety of "visionary encounters with departed loved ones." Mediums such as James Van Praagh, John Edward, and George Anderson have mediated contact with numerous deceased people by describing the impressions they receive from them.

ADCs have also been known to occur randomly and spontaneously, without anyone mediating them or guiding the experience. And now psychotherapists have learned to induce such experiences. Allan Botkin, a qualified psychotherapist, head of the Center for Grief and Traumatic Loss in Libertyville, Illinois, and colleagues claim to have successfully induced ADCs in nearly three thousand patients.

It appears that ADCs can be induced in about ninety-eight percent of the people who agree to try them. Usually the experience comes about rapidly, almost always in a single session. It is not limited or altered by the grief of the subject or his or her relationship to the deceased. It also does not matter what the experiencers believed prior to undergoing the experience; they could have been deeply religious, agnostic, or convinced atheists. ADCs can occur also in the absence of a personal relationship with the deceased—for example, in combat veterans who feel grief for an anonymous enemy soldier they had killed. And they can occur without guidance by the psychotherapist. Indeed, as Dr. Botkin reports, leading the experiencing subject actually inhibits the unfolding of the experience. It is sufficient that the therapist induces the mental state necessary for the experience to occur. This state is a slightly altered state of consciousness, brought about by means of a series of rapid eye movements. Known as "sensory desensitization and reprocessing," it produces a receptive state in which people are open to the impressions that appear in their consciousness.

Typically, the experience of after-death communication is clear, vivid, and thoroughly convincing. The therapists hear their patients describe communication with the deceased person, hear them insist that their reconnection is real, and watch repeatedly as their patients move almost instantly from an emotional state of grieving to a state of relief and elation.

MARK'S EXPERIENCE OF
AFTER-DEATH COMMUNICATION*

About twenty-five years ago Mark was embarking on a successful professional career when one night, driving alone, he was blinded by car lights and strayed into the path of an oncoming car. He was not injured, but the young family in the other car, father, mother, and twelve-year-old girl, were killed. Mark's life changed from that day; he awoke each morning to deep sadness and severe guilt, and plodded through the day reliving the accident over and over again. He twice attempted suicide, had two

*Reported in Botkin and Hogan, *Reconnections: The Induction of After-Death Communications in Clinical Practice.*

failed marriages, and was on the verge of losing his job. Life appeared to have ended for him. He then tried an ADC experience induced by Dr. Botkin. Following the brief interval of eye-movement desensitization and reprocessing, Mark sat quietly, with closed eyes. After a moment he said, "I can see them. It's the family with the little girl. They're standing together and smiling . . . Oh God, they look happy and peaceful. They're very happy being together and they're telling me they very much like where they are." Mark continued: "I can see each one very clearly, and especially the girl. She's standing in front of her mom and dad. She has red hair, freckles, and a wonderful smile. I can see the dad walking around, like he's showing me how he can walk. I have the feeling from him that he had multiple sclerosis before he died, and he is really happy he can now move around freely." Mark told the family that he is very sorry about what had happened and heard them say that they forgive him. He felt as if a huge burden had been lifted from him.

Mark had never actually seen the family; because of his deep grief and depression, he refused to look at pictures or read reports about them. But after the ADC experience he was feeling so much better that he stopped by his sister's house and looked at clippings of the accident. He said that he "freaked out." The newspaper pictures were very clearly of the same family he had experienced in his ADC, down to the smallest detail, such as the smile and the freckles of the girl. And there was a still more remarkable aspect: the father showing happily that he can walk. The newspapers reported that he indeed had multiple sclerosis at the time he died . . .

Mark's experience is fairly typical. In ADCs people experience the person they grieve for as happy and well, often younger than they were at the time they died. This "reconnection" with the deceased relieves and often fully resolves the grief weighing on the mind of the experiencer.

Clearly, ADCs have remarkable therapeutic value. But what do they mean? Are they grief-induced delusions? Botkin argues that they are not: they do not fit any of the known categories of hallucinations. If so, are they real: do the subjects actually encounter the deceased for whom they are grieving? That would suggest that the deceased still exists in some way, perhaps in another dimension of reality. This would be true immortality: the survival of the person after the physical demise of the body. This is a hopeful conclusion, but it is not likely to be true. There is another, more plausible, explanation and the informed universe can furnish it. It is simple and basic. At every moment throughout our life we read what we think, feel, and perceive into the A-field, a holographic field that preserves the experiences of our entire lifetime.

The A-field carries the holograms of our body and brain, and also carries the holograms of the communities in which we participate and of the milieu in which we live. Every element of these holograms can be individually retrieved by our brain. Retrieving the elements of our own hologram gives us the astonishingly complete and encompassing memory store that comes to light in near-death experiences and other altered states of consciousness. It extends to all things we have ever experienced in our lifetime, including our experience of the womb and of birth.

But this is not all: we can also read out the holograms of other people, and thereby relive their experiences. The people whose experiences we relive may be living or dead; the holograms in which their lifetime experiences are encoded do not phase out in time. As long as there are humans on this planet—and humanoid beings on other planets in the universe—the lived experiences of all people can be relived, over and over again.

When other people read out our own experience, *we* live again in *their* experience. When we read out other people's experience, *they* live again in *our* experience. And when we enter into communication with a person whom we grieve for, we do not communicate with that person directly, but read in the A-field the holograms created by his or her body and brain. These are complex, multiplex holograms that encompass the experience of an entire lifetime. We have seen that in altered states of consciousness people often communicate with the deceased they grieve for not as they were at the time of their death, but how they were earlier in life. This is possible and it stands to reason. Seeing one's loved ones as young and healthy is more conducive to alleviating and resolving one's grief than seeing them old and suffering.

The conclusion is evident. We as individuals are not immortal, but our experience is. The traces of everything we have ever experienced persist, and they can be forever recalled.

Prophets, philosophers, and spiritual people have often taken the traces we leave in the A-field as evidence for an immortal soul. Plato spoke of the immortality of the Soul, the aspect of the human being that springs from, and then returns to, the realm of eternal Forms or Ideas. Hegel considered the human mind the self-actualization of what he called the Absolute Idea through its temporal embodiments. Bishop Berkeley viewed the human mind as a reflection of the Divine Mind, the quintessence of the world's reality. Alice Bailey's intuitions match the latest insights from science remarkably: she located the source of human immortality in the "ether." "This word 'ether,'" she wrote, "is a generic term covering the ocean of energies which are all interrelated and which constitute that one synthetic energy body of our planet . . . the etheric or energy body, therefore, of every human being is an integral part of the etheric body of the planet itself."

Gustav Fechner, the pragmatic founder of experimental methods in psychology, expressed the same idea in surprisingly definite terms. "When one of us dies," he wrote after recovering from an illness, "it is as if an eye of the world were closed, for all perceptive contributions from that particular quarter cease. But the memories and conceptual relations that have spun themselves round the perceptions of that person remain in the larger Earth-life as distinct as ever, and form new relations and grow and develop throughout all the future, in the same way in which our own distinct objects of thought, once stored in memory, form new relations and develop throughout our whole finite life."

Nothing in this world is evanescent; all things continue to exist through the traces they leave in the cosmic information field. We humans, too, create an Akashic record of our lifetime experiences, a record that can be retrieved by others. Our individual experience is not limited to ourselves and to our individual lifetime. It can be reexperienced and thus relived at any time and at any place, today and at all times in the future.

Reincarnation

Understanding that it is the A-field—the information field of the cosmos—that confers immortality on us and not an individual immortal soul gives us a different perspective on reincarnation. This perspective is fully consistent with the evidence we have of reincarnation, which consists of impressions and ideas recounted by people about sites, people, and events they have not and could not have encountered in their present lifetime. It is then assumed that they encountered them in previous lifetimes. These "past-life experiences" have an element of truth in them, but that does not guarantee that such experiences come truly from a past life.

"Past-life stories" crop up routinely in the experience of psychotherapists who practice regression analysis. They place their patients in a slightly altered state—hypnosis is not needed, since breathing exercises, rapid eye movements, or simple suggestion is usually sufficient—and take them back from their current experiences to the experiences of their past. They can often move their patients back to early childhood, infancy, and physical birth. Experiences that seem to be those of gestation in the womb surface as well.

Interestingly, and at first quite unexpectedly, psychotherapists have found that they can take their patients back further than the womb and physical birth. After an interval of apparent darkness and stillness, other experiences appear. They are of other places and other times. Yet the patients not only recount them as the experience of a novel they have read or a film they have seen, but actually relive them as well. As Stanislav Grof's records testify, they *become* the person they experience, even to the inflection of voice, the language (which may be one the patient has never known in his or her present lifetime), and, if the experience is of infancy, the involuntary muscle reflexes that characterize infants.

Ian Stevenson, of the University of South Carolina, investigated the past-life experiences recounted by children. During more than three decades Stevenson interviewed thousands of children, in both the West and the East. He found that from the age of two or three, when they begin to verbalize their impressions, until the age of five or six, many children report identification with people they have not seen, heard of, or encountered in their young lives. Often these reports can be verified as the experience of a person who had lived previously, and whose death matches impressions reported by the child. Sometimes the child carries a birthmark that is associated with the death of the person with whom she or he identifies—such as an indentation or discoloration on the part of the body where a fatal bullet entered, or malformations on the hand or foot the deceased had lost or had wounded.

The experiences reported by children—and by grown-ups in altered states of consciousness—actually occur, and they show that we can access the experiences of other people whether they stand before us or are far away, and whether they are living today or have lived sometime in the past. But when we reexperience other people's experiences we do not reincarnate them, for the images and ideas that surface in our consciousness stem not from single individuals whose soul has survived their death and is now reincarnated in us. Rather, the ideas, images, and impressions entering our consciousness have their source in the vacuum. The information carried in the vacuum's A-field is active and effective. Its range is vast; it embraces other humans as well as other forms of life, and all things in the universe. In integrating with it, it is not our individual body and our individual soul, but our individual *experience* that achieves immortality.

We do not disappear from the world without a trace; all that we experience becomes part of the collective memory bank of humankind, to be read out again and again. We can live on in the brain and consciousness of people today, and in all future generations.

NINE

The Poetry of Cosmic Vision

At the cutting edge of the sciences, a new concept of the world is emerging. In this concept all things in the world are recorded and all things inform one another. This gives us the most encompassing vision we have ever had of nature, life, and consciousness. It gives us an integral theory of everything.

In this concluding chapter, we apprehend the new world concept not as a rationally argued scientific theory, but as a poetic vision that conveys its spontaneous *feel*. This is important. If the rediscovered information- and memory-filled universe is the best insight we have ever had into the nature of reality, we should know it not only with our rational faculties; we should apprehend it also with our creative imagination. We should not just grasp it with our intellect, but also feel it in our hearts and in our guts.

Here, then, is a vision that is *imaginative* but not *imaginary:* the feeling-portrait of the universe that is now emerging at the frontiers of the sciences.

Imagine, if you will, a lightless, soundless, formless plenum. It is filled both with the primeval consciousness that is the womb of all mind and spirit in the cosmos and with the fluctuating energies out of which all things come to exist in space and in time. There is no-thing in this cosmic fullness, yet there is every-thing, in potential. Everything that can and will ever happen is here, in formless, soundless, lightless, quiescent turbulence.

After an infinity of cosmic eons, a sudden explosion, untold mag-nitudes greater than any turbulence ever witnessed or even imagined by human beings, penetrates the formless turbulence; a shaft of light rises from its epicenter. The plenum is no longer quiescent; it is rent by a supercosmic force emerging from its hitherto soundless and lightless depth. It liberates gigantic forces, transforming the plenum from virtual formlessness into dynamic formative process. The surface foams with instantly appearing and disappearing ripples of energy, forming and annihilating in a cosmic dance of unimaginable speed and momentum. Then the initial demented rhythm becomes more sedate, the foam more orderly. The ripples radiate outward from the epicenter, bathed in pure light of infinite intensity.

As the foam expands, it becomes grainier. Swirls and vortexes appear, incipient if as yet evanescent wave-patterns modulating the sur-face of the evolving plenum. With the passing of further cosmic eons, the ripples of patterned energy consolidate into lasting forms and structures. They are not separate from each other, for they are micro-patterns struc-turing into larger patterns within a common wave-field. They are part of the underlying and now no longer formless plenum that erupted and cre-ated them. Each ripple is a microworld in itself, pulsating with the lib-erated energies of the plenum and reflecting in its micrototality the macrototality from which it emerged.

The micro-patterns trace their careers in the expanding space of the initial explosion and take on structure and complexity. They modulate the turbulent plenum. It is more and more structured at the surface, as the ripples cohere into complex wave-structures; and it is more and more modulated below, as the evolving structures create minute vortices that integrate into information-carrying holograms. The informed holofield below and the micro-patterns on the surface evolve together. Their growing architecture enriches the holofield, and the enriched holofield in-forms the evolving micro-structures. Surface and depth coevolve, tak-ing on complexity and coherence.

The more complex the structures that emerge, the more independ-ent they appear of the depth below. Yet the ripples and waves at the surface are not separate but part of the medium from which they arise—they are like "solitons," the curiously object-like waves that emerge in a turbulent medium.

The ripples and waves cohere in elaborate structures, subtly interconnected with each other. At a crucial stage of their evolution they become self-sustaining, reproducing themselves and replenishing spent energies from the embedding energy fields.

The evolving wave-patterns have not just external relations; they also have an inner reflection: a "feel" of each other and of the depth. At first an unarticulated basic sensation, this inner reflection gains in articulation as the self-maintaining waves acquire structure and complexity. They develop higher and higher grades of inner reflection, articulating their basic feel of the world as a representation of individual things and processes. They map the world that envelops them, and themselves in that world.

After another cosmic eon, the energies liberated by the initial explosion dissipate across the surface of the plenum. Some mega-structures use up the free energies available to them and explode, strewing their micro-ripples into space where they consolidate into new mega-structures. Others implode, and in a final flash reenter the plenum from which they emerged. The ripples that evolve on the surface of smaller mega-structures break down, incapable of maintaining themselves in an environment of fading energy. As the universe ages, all complex structures and articulated reflections disappear. But although the surface loses modulation, the memory of the depth is not affected: the holograms created by the ripples remain untouched. They conserve the trace of the surface's evanescent structures together with their feels and reflections.

And now another shaft of light rends the plenum, breaking its quiescent turbulence and reviving it with another formative burst: a new universe is born. This time the ripples and structures that form on the surface do not appear randomly, at the mercy of chance: they derive from a plenum in-formed with the holo-trace of prior ripples and waves.

The cosmic drama repeats time after time. Further shafts of light radiate outward from the epicenter, another multitude of ripples moves outward to dance, to cohere, to feel, and to reflect. The new universe ends as the ripples and the structures it brought into being vanish at the surface. But the holograms created by them in the depth inform the next universe, born as further explosions rend the plenum. Time after time, the cosmic drama repeats, but it does not repeat in the same way. It builds on its own past, on the memory of the ripples and waves that appeared and then disappeared in prior universes.

In universe after universe the plenum brings forth micro-ripples and mega-wave stuctures. In each universe the ripples and waves vanish, but their memory lives on. In the next universe new and more elaborate structures appear, with more articulated reflections of the world around them.

In the course of innumerable universes, the pulsating Metaverse realizes all that the primeval plenum held in potential. The plenum is no longer formless: its surface is of unimaginable complexity and coherence; its depth is fully informed. The cosmic proto-consciousness that endowed the primeval plenum with its universe-creative potentials becomes a fully articulated cosmic consciousness—it becomes, and thenceforth eternally is, the self-realized mind of God.

An Autobiographical
Retrospective

*Forty Years in Quest of
the Integral Theory of Everything*

Science and the Akashic Field is the product of four decades of searching for meaning through science. I started on this quest in the spring of
1959, shortly after my first son was born. Until then my interest in
philosophical and scientific questions had been just a hobby—I had
been traveling the world as a musician, and nobody, not even I, had
ever suspected that it would become more than an intellectual pastime.
But my interest in finding a meaningful and encompassing answer to
what I experienced and knew about life and the universe grew, and the
quest that began in 1959 became an all-consuming vocation. It culmi-
nated four decades later in the spring of 2001, as I sat down to draft
out *The Connectivity Hypothesis,* my latest theoretical work. The present book, summarizing my findings for the general readership, followed
in 2002–2004.

My enduring interest has been to find an answer to questions such
as "What is the nature of the world?" and "What is the meaning of my
life in the world?" These are typically philosophical questions—
although the majority of today's academic philosophers prefer to hand
them to theologians and poets—yet I did not seek to answer them
through theoretical philosophy. While I was not an experimental scientist (and given my background and interest I was not attempting to

become one), I did have a strong sense that the best way to tackle these questions is through science. Why? Simply because empirical science is the human endeavor that is the most rigorously and systematically oriented toward finding the truth about the world, and testing its findings against observation and experience. I wanted the most reliable kind of answers there are, and reflected that I could find no better source for them than science.

For a young man in his twenties without formal background in a specific field of science, this was quite presumptuous. I would like to call what I had intellectual courage, but at the time I did not feel especially courageous—just curious and committed. Nonetheless, I was not entirely unprepared, for I had done a good deal of prior reading (mostly on planes and trains and in hotel rooms) and took part in various college and university courses. Being a successful concert pianist, I never enrolled for an academic degree for which I saw no conceivable use.

In 1959 I turned over a new leaf: I set about doing systematic reading and research. What was until then a favorite hobby became a methodical quest. I started with the foundations of science in classical Greek thought and moved to the founders of modern science before turning to contemporary science. I was interested neither in the technical details that take up the lion's share of the training of science professionals—techniques of research, observation, and experimentation—nor in controversies about methodological or historical fine points. I wanted to get straight to the heart of the matter: to find out what a given science could tell me about the segment of nature it investigates. This required a good deal of spadework. The findings were unexpectedly sparse, consisting of a few concepts and statements, usually at the end of extensive mathematical and methodological treatises. They were, however, extremely valuable, much like nuggets of gold that come to hand after sifting through streams of water and mountains of ore.

In the course of the 1960s, I learned to do my sifting rapidly and efficiently, covering a good deal of ground. What meaning I found half-buried in particular fields I jotted down, and attempted to bring it into relation with what I found in other fields. I did not intend to write a treatise or create a theory, I just wanted to understand what the world and life—my life, and life in general—are all about. I made copious notes, but never expected that they would get into print. How they did so is one of the curious episodes of my life.

After a successful concert in The Hague, I found myself sitting at late supper next to a Dutchman who brought up some of the very questions that fascinated me. I got into conversation with him, and ended by going up to my hotel room to show him the notes I always had with me. He retired into a corner and began reading. Shortly after that he disappeared. I was concerned, since I had no copy. However, the next morning my newfound friend reappeared with my notes under his arm. He announced that he wanted to publish them. This was a surprise, for I knew neither that he was a publisher (he turned out to be the philosophy editor at the renowned Dutch publishing house Martinus Nijhoff) nor that my notes would merit publication. Of course, they required a good deal of completing and organizing before they could be published in book form. But published they were, a year and a half later (*Essential Society: An Ontological Reconstruction*, 1963).

The experience in The Hague reinforced my determination to pursue my quest. I joined the Institute of East European Studies at Switzerland's University of Fribourg, and for several years combined writing and research with concert work. I came out with another, less theoretical, book shortly after the first (*Individualism, Collectivism, and Political Power*, 1963) and a few years later published another philosophical treatise (*Beyond Scepticism and Realism*, 1966). The period of writing and researching combined with concertizing came to an end when, in 1966, I received an invitation from Yale University's Department of Philosophy to spend a semester there as visiting fellow. Accepting that invitation was a major decision, for it meant exchanging the concert stage for the life of an academic.

The decision to go to Yale—which led to teaching appointments at various U.S. universities and, in 1969, to a Ph.D. at the Sorbonne in Paris—gave me the opportunity to pursue my quest full time. Although in any established university there is considerable pressure to keep to the rather narrowly defined territory of one's own field, I never wavered from the conviction that there is meaning to be discovered in the world at large, and that the best way of discovering it is to query the theories put forward by leading scientists in all the relevant fields, not just those that belong to one's area of specialization. I was fortunate in finding colleagues—first at Yale, then at the State University of New York—who understood this conviction and helped me overcome the academic hurdles that would have stood in the way.

The search for meaning through science called for considerable time and energy. I soon realized that, like Archimedes, I needed firm ground from which to start. I found two basic alternatives. One was to start with the stream of one's own conscious experience and see what kind of world one could logically derive from that experience. The other was to gather all the information one can about the world at large, and then see if one can account for one's own experience as the experience of that world. The former has been the method of the empirical schools of Anglo-Saxon philosophy and of that branch of continental philosophy that took its cue from Descartes, and the latter the method of naturalistic metaphysics and science-based philosophy. I read up on these schools, paying special attention to Bertrand Russell and Alfred Ayer among the British philosophers, Edmund Husserl and the phenomenologists of the continental schools, and Henri Bergson and Alfred North Whitehead among the naturalistic process philosophers. I concluded that neither the formal analysis of experience nor the introspective method of the phenomenologists leads to a meaningful concept of the real world. These schools ultimately get bogged down in what philosophers call the "ego-centric predicament." It appears that the more systematically one investigates one's immediate experience, the less easy it is to get beyond it to the world to which that experience presumably refers. We are logically obliged to take the initial leap of assuming the objective existence of the external world, and then to create the scheme in light of which our experience makes sense as the human experience of that world.

In *Beyond Scepticism and Realism,* I contrasted the "inferential" approach that starts from one's own experience with the alternative "hypothetico-deductive" method that envisages the nature of the world and explores how our observations accord with it. I concluded that, ideally, the overlap between these distinct and sometimes seemingly contradictory approaches is what gives the most reliable information about the real nature of the world. I identified some areas of overlap, but did not stop there: I wanted to get on with my quest, and began to explore the bold hypothetico-deductive approach. To my considerable relief, I found that this approach had been adopted by many great philosophers and practically all theoretical scientists, from Newton and Leibniz to Einstein and Eddington.

Einstein stated the principal premise of the naturalistic approach. "We are seeking," he said, "for the simplest possible scheme of thought that will bind together the observed facts." The simplest possible scheme, I realized, cannot be inferred from observations: as Einstein said, it needs to be imaginatively envisaged. One must search for and codify the relevant observations, but one cannot stop there. While empirical research is necessary, the creative task of putting together the resulting data in ways that they make sense as meaningful elements of a coherent system cannot be neglected. It is the principal challenge facing the inquiring mind. The attempt to "create the simplest possible scheme of thought that will bind together the observed facts" (and by "observed facts" I meant all the facts needed to make sense of the world) defined my intellectual agenda for the next four decades.

The scheme I first envisaged rested on the organic metaphysics of Whitehead. In this conception, which dated originally from the 1920s, the world and all things in it are integrated and interacting "actual entities" and "societies of actual entities." Reality is fundamentally organic, so living organisms are but one variety of the organic unity that emerges in the domains of nature. My subsequent readings in cosmology and biology confirmed the soundness of this assumption. Life, and the cosmos as a whole, evolves as integrated entities within a network of constant formative interaction. Each thing not only "is," it also "becomes." Reality, to cite Whitehead, is process, and an integrative evolutionary process at that.

The question I asked was how I could identify the evolving entities of the world in such a way that they would make sense as elements in an organically integral universe. Colleagues at Yale called my attention to the work of Ludwig von Bertalanffy in the area of "general system theory." Bertalanffy was attempting to integrate the field of biology in an overall scheme that would lend itself to further integration with other domains of natural science, and even with the human and social sciences. His key concept was "system," conceived as a basic entity in the world. Systems, he argued, appear in similar ("isomorphic") ways in physical nature, living nature, as well as the human world. This was most helpful: it supplied the conceptual tool I was looking for. I read Bertalanffy, then met with him and developed the concept of what we jointly decided to call "systems philosophy."

Introduction to Systems Philosophy (1972) was a painstakingly researched book—it took five years to write—and when it was published I was tempted to rest for a while on my laurels. But I was not satisfied. I needed to find an answer in leading-edge science not only to how systems are constituted and how they relate to each other, but also to how they change and evolve. Whitehead's metaphysics gave me the general principles and Bertalanffy's general system theory clarified the relations between systems and environments. What I still needed was the key to understanding how these relations can lead to integrative and on the whole irreversible evolution in the biosphere, and in the universe as a whole.

To my surprise, the key was furnished by a discipline about which I knew little at the time: nonequilibrium thermodynamics. I reached this conclusion on the basis of my brief but intense friendship with Erich Jantsch, who died unexpectedly a few years later. He directed my attention to the work, and subsequently to the person, of the Russian-born Nobel laureate thermodynamicist Ilya Prigogine. The latter's concept of "dissipative structures" that are subject to periodic "bifurcations" furnished the evolutionary dynamic I needed. After discussing this concept with Prigogine, my work focused on what I called "general evolution theory." The basic kind of entity that populates the world transformed in my thinking from Whitehead's "organism" and Bertalanffy's "general system" to Prigogine's nonlinearly bifurcating "dissipative structure," an evolving thermodynamically open system. The world began to make more and more sense.

Apparently, the sense I made of the world also intrigued scholars in fields other than systems theory and philosophy. While teaching and researching at the State University of New York at Geneseo, to my surprise I received a phone call from Richard Falk, of Princeton University's Center of International Studies. Falk, one of the foremost "world system" theorists of the time, asked me to come to Princeton to lead a series of seminars on the application of my systems theory to the study of the international system. I assured him that I knew next to nothing about the international system and had only vague notions of how my theory would apply to it. But Falk was not to be deterred. He and his colleagues, he said, would see to the application of my theory if I would come and discuss that theory with them. This I agreed to do.

The experience of my Princeton seminars was intellectually rewarding as well as exciting: it opened new vistas. I found a new and intensely practical application for general system theory, systems philosophy, and general evolution theory: human society and civilization. Society and civilization, I realized in the mid-1970s, are undergoing a process of irreversible transformation. The human world is growing beyond the bounds of the nation-state system to the limits of the globe and the biosphere. This called for rethinking some of our most cherished notions about how societies are structured, how they operate, and how they develop. With valuable input from Richard Falk and other Princeton colleagues, I spelled out my evolutionary conception of the world system in *A Strategy for the Future: The Systems Approach to World Order* (1974).

Strategy elicited attention beyond academia. Another call followed, this time from Aurelio Peccei, the visionary Italian industrialist who founded the world-renowned think tank known as the Club of Rome.

He suggested that I apply the systems approach to the "limits to growth" problem, focusing not on the limits themselves (as Jay Forrester and Dennis and Donella Meadows did in the first report to the club, *The Limits to Growth*), but on the ambitions and motivations that drive people and societies to encounter the limits. This invitation was an intellectual challenge with major practical relevance—it could not be refused. I took a leave of absence from my university and moved to the UN headquarters in New York. Davidson Nicol, executive director of the UN's Institute of Training and Research (UNITAR), invited me to join his institute in order to create the international team that was to work on this project. Within a year, some 130 investigators on six continents were enlisted in creating the Club of Rome's third report, focusing on humankind's "inner" rather than "outer" limits (*Goals for Mankind: The New Horizons of Global Community,* 1977).

Having finished the report, I repaired to my university to resume researching, writing, and teaching. This, however, was not to be. A further call from Nicol asked me to represent UNITAR at the founding of the United Nations University in Tokyo, and when I filed my report Nicol asked me to stay on at the institute to head research on the hottest subject of the day, the "new international economic order." This was another challenge that could not be ignored. After three years of intense work, fifteen volumes written with collaborators from ninety research institutes in every part of the world were published in a series created for this purpose by Pergamon Press of Oxford: the New International Economic Order Library. The NIEO Library was to produce background documentation for the General Assembly's landmark General Session of 1980, which was to launch the "global dialogue" between the developing South and the industrialized North. But the big powers of the North refused to enter the dialogue and the UN system dropped the project of the new international economic order.

When I was about to return to my university to pursue at last my principal quest, UN Secretary-General Kurt Waldheim asked me to suggest other ways in which North–South cooperation could be pursued. The proposal I made to him and to UNITAR was based on systems theory: it was to insert another "systems level" between the level of individual states and the level of the United Nations. This was the level of regional social and economic groupings. The project, called Regional and Interregional Cooperation, was adopted by UNITAR and took four years of intense work to carry out. In 1984 I reported the results in four bulky volumes that accompanied a declaration of a specially convened "panel of eminent persons." Due to internal politics, the declaration was not handed to the secretary-general and thus could not be made into an official document, but its text was circulated to all member-state delegations. Disappointed with this outcome but hopeful that sooner or later the proposals contained in the declaration would bear fruit, I decided that I had merited a sabbatical year. I moved with my family to our converted farmhouse in Tuscany. That sabbatical year, begun in 1982, has not yet come to an end.

However, the 1980s and '90s turned out to be much more than a "read and write" sabbatical. It was a time of increasingly intense international commitments. In the 1980s I was involved in discussions at the Club of Rome, then took a major part in the United Nations University's European Perspectives project. Subsequently, I served as science adviser to Federico Mayor, the two-term director-general of UNESCO. But since 1993 the brunt of my attention was focused on the Club of Budapest, the international think-tank I founded that year to do what I had hoped the Club of Rome would do: center attention on the evolution of human values and consciousness as the crucial factors in changing course—from a race toward degradation, polarization, and disaster to a rethinking of values and priorities so as to navigate today's transformation in the direction of humanism, ethics, and global sustainability. As reports to the Club of Budapest I wrote *Third Millennium: The Challenge and the Vision* (1997) and most recently *You Can Change the World: The Global Citizen's Handbook for Living on Planet Earth* (2003).

Notwithstanding these activities and commitments, I remained faithful to my basic quest. When in 1984 I left the United Nations for the Tuscan hills, I took stock of how far I had gotten. I found that I needed to go further. Systems theory, even with the Prigoginian dynamic, provided a sophisticated but basically local explanation of how things relate and evolve in the world. The open system dynamic of evolution refers to particular systems; their interaction with other systems and the environment constitutes what Whitehead termed "external" relations. Yet Whitehead affirmed that in the real world all relations are *internal:* every "actual entity" is what it is because of its relations to all other actual entities. With this in mind, I set about reviewing the latest findings in quantum physics, evolutionary biology, cosmology, and consciousness research, and found that the idea of internal relations is entirely sound. Things in the real world are indeed strongly—"internally," "intrinsically," and even "nonlocally"— connected and correlated with each other.

Internal relations also bind our own consciousness with the consciousness of others. This was brought home to me by a personal experience that I recounted in 1993 in the preface to *Creative Cosmos* and will not repeat here. Although a mystical experience does not provide proof of internal relations between one's mind and the mind of others, it does provide an incentive to study the possibility that such relations exist. This consideration became part of my explorations in the years that followed.

The science books I have produced in this "Tuscan period" are—in addition to the book in the hands of the reader—*The Creative Cosmos* (1993), *The Interconnected Universe* (1995), *The Whispering Pond* (1997–98), and *The Connectivity Hypothesis* (2003). In these books I marshal evidence that things in the real world are intrinsically interconnected, and suggest the reason for it. The theory of the information field—which I first called the psi-field and am now calling the "A"(for Akashic)-field—provides that reason. It claims that the connections and correlations that come to light in the physical and the life sciences, the same as the transpersonal ties that emerge in experimental parapsychology and consciousness research, have one and the same root: a subtle but entirely fundamental coherence- and correlation-creating field at the heart of the universe. Further clarifying and codifying the nature and effects of this field would be of the utmost importance. It would bring science significantly closer to Einstein's (and to my) ultimate goal of finding the "simplest possible scheme that can bind together the observed facts."

My recent books, culminating with *Connectivity Hypothesis* and *Science and the Akashic Field*, state, I believe, the essential framework for the simplest possible scheme that can bind together the remarkable facts that are now coming to light at the cutting edge of the sciences.

THE AUTHOR'S JOURNEY MIRRORED IN COMMENTS BY SOME OF THE FOREMOST SCIENTISTS AND THINKERS OF OUR TIME

Ludwig von Bertalanffy on systems philosophy:

"Laszlo's pioneering work develops systems philosophy both in breadth and depth. As he argues convincingly, contemporary 'analytic' philosophy is in danger of 'analyzing itself out of existence' . . . What we need, says Laszlo, is rather a 'synthetic' philosophy, that is, one which receives new inputs from the various developments in modern science and tries to follow the other way in philosophy, namely, endeavors to put together the precious pieces of specialized knowledge into a coherent picture. . . ."

"Laszlo's work is the first comprehensive treatise of 'systems philosophy.' No one who looks beyond his own specialty and narrow interests will be able to deny the legitimacy of this quest."

FOREWORD TO *INTRODUCTION TO SYSTEMS PHILOSOPHY* (1972)

Richard Falk on systems theory applied to the world system:

"We cannot be optimistic about the future of the human species unless we envisage a rather drastic restructuring of social, economic, and political life on the planet. . . . One encouraging development is the increasing number of serious efforts . . . to find the means to build the sort of world society that has the capacity to deal with the problems confronting humankind. Among these intellectual efforts none is more significant than the work of Ervin Laszlo, who has brought to bear the well-developed framework of general systems analysis on the specific task of constructing a just and viable system of world order. In *A Strategy for the Future*, Laszlo portrays with intellectual power and originality the contours of a desirable world system and provides a carefully interrelated concept of how we get from here to there."

". . . world-order studies are, I think, with Laszlo's help being liberated from their literary and sentimental origins and achieving the status of a new academic discipline of normative content that deals with evidence, explanation, and prediction. . . . What Laszlo provides . . . is a framework based on systems theory that can accommodate information drawn from any discipline or perspective and an insistence that the future of the human race is too important to be left in the hands of statesmen, generals, cartelists, and the like—who, in any event, are disastrously confined by the predispositions and interest structure of the state system."

". . . I believe Laszlo has put us on the right track in an innovative and exciting way. His leadership in the systems area is itself one element in a new movement for global reform taking place among intellectuals throughout the world. In my view, anyone concerned with the future of humankind and eager to take part in its creation has a special obligation to read what Laszlo has written. His book deserves to be one of the main texts for the reeducation of the mind that must occur if we are ever to become both good citizens and good people."

INTRODUCTION TO *A STRATEGY FOR THE FUTURE* (1974)

Jonas Salk on general evolution theory:

"In this book Ervin Laszlo has turned his integrating mind to the task of bringing together observations that reveal the operation of the laws of nature in evolving emergent systems of increasing complexity. . . . The grand sweep of evolution over the expanse of time that has thus far elapsed is revealed in this book in a form that is useful for the nonscientist and the scientist alike."

"There is emerging a new literature on the subject of evolution, one that has expanded far beyond the limits of the work of Darwin and Wallace, who first made us aware of evolution in the origin of species. Since then evolution has come to be seen in a wider context. It is now seen in its universality, in its universal presence, and in its absence, as when species cease to evolve and are no longer able to persist. We now see the meaning of this in the human realm with the emergence of the capacity to evolve as the most valuable of all human attributes."

"It is for this reason that we need to understand evolution fundamentally if we are to be able to maintain our place in the evolutionary scheme of things as an evolving species. . . . This book will help make us aware of the awesome challenge that this turn of events presents to us and to future generations. Can we rise to it? Time will tell. Do we have enough time? I presume that we do, provided we don't waste it. That's the meaning of the surge in interest in evolution in our time to which this useful, comprehensive, and illuminating book is a response."

FOREWORD TO *EVOLUTION:*
THE GRAND SYNTHESIS (1987)

Ilya Prigogine on systems and evolution theory applied to the contemporary world:

". . . Laszlo's study [*The Age of Bifurcation*] represents a remarkable coincidence: at this very moment, mankind is living through a crucial time of transformation while science is undergoing a spectacular transition. More and more, an ever-increasing number of scientists perceive that a new paradigm is taking shape. Everywhere we see fluctuations, evolution, diversification. This is true not only on the level of macroscopic phenomena, as in chemistry, but also on the microscopic level in particle physics and on the vast scale of modern cosmology."

"The title of this book, *The Age of Bifurcation,* is well chosen because with the concept of bifurcation the historical category of 'event' enters contemporary science. An event is something that cannot be deterministically predicated. The position of the Earth around the Sun in a given number of years could hardly be considered an event, while obviously the birth of Mozart was an event in the history of Western music."

". . . we now have hope that with our achievements, both theoretical and experimental, with our immensely improved capacity of producing wealth, and with our new facilities for interpersonal communication, we can come at last to a form of civilization where an increasing number of people have the possibility to manifest the creativity which, I believe, is present in every human being. Is this the beginning of such a new age? We are still too heavily involved in the planetary transformation underway to reach a firm evaluation, but perhaps—and this is my hope—succeeding generations will see our time as the beginning of a great age of bifurcation—and will look upon this book as the herald to that age."

FOREWORD TO *THE AGE OF BIFURCATION:*
UNDERSTANDING THE CHANGING WORLD (1991)

Arne Naess on the holistic theory of the A-field (also called the quantum/vacuum interaction hypothesis):

"The creative work of Ervin Laszlo is a brilliant testimony of how conceptual imagination—deductively related to careful observation—can make us see the cosmos, and our place within the cosmos, in new ways that are of great inspirational value. Reality as conceived by Ervin Laszlo has what I call 'gestalt character'—a predominance of internal rather than external relations."

"A central part of Laszlo's conceptual framework is the quantum/vacuum interaction (QVI) hypothesis. This is a highly sophisticated theory, rather than a hypothesis, in my terminology. . . . Greatly simplified, one might say that Laszlo envisions a world that is constantly created, and here every event that happens locally, even an event in one's consciousness, is connected with events that happen everywhere else."

"There are many of us in science and philosophy who wish to see a growing trend of bold theory formulation inspired by such courageous yet unpretentious efforts as the present study by Ervin Laszlo."

FOREWORD TO *THE INTERCONNECTED UNIVERSE* (1995)

Karl Pribram on the holistic theory of the A-field (also called the quantum/vacuum interaction hypothesis):

"*The Creative Cosmos* is a superb example of postmodern deconstruction at its very best. Its first two parts demonstrate the anomalies and lacunae in the current narrative we call science. The next sections boldly develop a new narrative that aims to carry our comprehension beyond these limitations. . . .The narrative aspects of science, the concepts and meanings to which the computations point, have been neglected, often deliberately as in the ever-popular Copenhagen interpretation of quantum physics. This neglect has produced considerable malaise in some of us, and more important, it has led to a cover-up of the anomalies and lacunae addressed in *The Creative Cosmos*. [This book] ably summarizes what is missing in today's account of science-as-narrative. Of course, Laszlo is not alone in his lament. Einstein, Dirac, Bohm and Bell have all attempted to understand their formulations in physics; Koestler, in biology and psychology. But the received wisdom in the classroom has, for the most part, emphasized the elegance of what has been achieved often with the advice that any attempt at further understanding would simply confuse."

"Laszlo is to be commended in that he provides us with a plausible alternative. All of the scientists noted above have groped in the direction now taken by Laszlo. He points out, that as the twentieth century comes to a close, scientists are again becoming more comfortable with the concept of 'field', which has been eclipsed for most of the century by an almost exclusive emphasis on the particulate."

"Gravitation, electromagnetic, the strong and weak nuclear forces have all become relatively familiar, at least to scientists, because their inferred properties do not invoke any radical departure from the measurements that have served scientists so well. . . . [T]he postulated fifth field [the A-field] is different. It is not inferred from an interaction among spatially and temporally separated entities. As Bohm has described it, space and time become implicate, enfolded. Mathematically, the fifth field is spectrally, holographically organized. The organization is composed of interference patterns, that is, of the amplitudes (amounts) of energy present at intersections among waveforms. . . . The fifth field is thus not a simple inference from observations. Rather, the fifth field is a transformation of fields which are inferred from observations."

"Laszlo has, indeed, filled the need for a twenty-first century renewal of the narrative of science which has been so neglected during the twentieth century."

FOREWORD TO *THE CREATIVE COSMOS* (1993)

Karan Singh on the holistic theory of the A-field (also called the quantum/vacuum interaction hypothesis):

"Perhaps the most significant development in recent times which, though the subject of several important books, has still not received the attention it deserves, is the growing convergence between the mystical worldview (predominantly, but by no means exclusively, Eastern) and the emerging paradigm of reality among scientists at the cutting edge of contemporary knowledge. *The Whispering Pond,* the latest in Ervin Laszlo's important series mapping out the geography of reality, makes this point, and does much to rectify it."

"With astonishing incisiveness and clarity, *The Whispering Pond* propounds a breathtaking vision. Its most significant upshot is that the scenarios of cosmic fate are likely to be open; fate and destiny are not sealed, and the future may not only happen but could also be created."

"In the light of the globalization of human civilization taking place before our very eyes, the evolution of a global consciousness is urgently needed if mankind is not to destroy itself and all life on this planet by its inability to responsibly manage its technological ingenuity. For such a global consciousness to arise, a worldview in which science and spirituality converge is a necessary development. The publication of *The Whispering Pond* is a significant step in this direction."

FOREWORD TO *THE WHISPERING POND* (1996)

David Loye on the holistic theory of the A-field (also called the quantum/vacuum interaction hypothesis):

"*The Whispering Pond* is an enormous contribution to our understanding at a critical time in human evolution. It gives us the vital new fragments of emergent 'truth' in language we can understand. And it provides the even more vital sense of the meaningful whole into which these fragments fit, which we have been lacking. This book, and the pioneering scientific study it is based on—Laszlo's *The Interconnected Universe*—call[s] to mind that watershed statement of the 18th century, *The Critique of Pure Reason*. There a philosopher with a similarly amazing capacity for integration—Immanuel Kant—so transcended in his synthesis the science and philosophy of his time as to establish a new framework for practically all of modern thought. It will be interesting to see if history repeats itself."

FOREWORD TO *THE WHISPERING POND* (1996)

Ken Wilber on today's revolution of consciousness:

"Ervin Laszlo can only be called a genius of systems thinking. In books too numerous to mention—my favorites include *The Systems View of the World, Evolution: The Grand Synthesis, The Choice, The Whispering Pond,* and *Third Millennium*—Ervin Laszlo has, probably more than any person alive, intricately spelled out a staggering but often neglected fact: we live in a hopelessly interconnected universe, with each and every single thing connected in almost miraculous ways to each and every other. His work, spanning four decades, has been a clear and consistent call to recognize the richly interwoven tapestry that constitutes our world, our lives, our hopes and our dreams. By rising to a vision of the whole, he has helped countless individuals escape the narrow limitations and depressing fragments that have haunted the modern world for at least three centuries."

FOREWORD TO *THE CONSCIOUSNESS REVOLUTION* (1999)

Ralph Abraham on The Connectivity Hypothesis:

"The new science of life of Sheldrake tries to restore vitalism to biology. The archetypal psychology of Jung, Hillman, and Moore tries to bring it back into psychology. Along with many others, these efforts may be seen as a New Renaissance. Amid the milieu of this paradigm shift, Ervin Laszlo stands out as the unique champion of a holistic philosophy of the broadest perspective. For his bold plan is to unify all—quantum, cosmos, life, and consciousness—in a single grand unified model. When a great grand unified theory will appear it will very likely conform to the prophetic vision of Ervin Laszlo."

FOREWORD TO *THE CONNECTIVITY HYPOTHESIS* (2003)

Christian de Quincey on The Connectivity Hypothesis:

"Laszlo has put together a remarkable summary of some of the latest findings in sciences such as quantum mechanics, cosmology, neuro-science, and consciousness studies, along with his renowned expertise in systems and complexity theory. He has woven together key elements from each of these sciences to make one of the most coherent cases yet for a radically different worldview based on the subquantum domain of the zero-point energy field, or what he calls 'the psi (A-field) field.' "

COMMENT ON *THE CONNECTIVITY HYPOTHESIS*

Stanislav Grof on The Connectivity Hypothesis:

"This is a brilliant summary of the major conceptual challenges for the Cartesian-Newtonian paradigm, which has dominated Western scientific thinking for the last three centuries. Laszlo outlines the areas in quantum physics, astrophysics, biology, and psychology where these disciplines encountered observations that they could not account for. But he does not stop there; he offers an elegant interdisciplinary model that could help to reconcile the existing paradoxes. Ervin Laszlo is a world-class scientist and his contributions are groundbreaking."

COMMENT ON *THE CONNECTIVITY HYPOTHESIS*

References
and Further Reading

A more detailed bibliography, including technical science papers, is given in the following books by the author:

The Creative Cosmos. Edinburgh: Floris Books, 1993.

The Interconnected Universe. Singapore and London: World Scientific, 1995.

The Whispering Pond. Rockport, Shaftesbury, and Brisbane: Element Books, 1996.

The Connectivity Hypothesis. Albany: State University of New York Press, 2003.

CHAPTER 1

Bateson, Gregory. *Steps to an Ecology of Mind*. New York: Ballantine, 1972.

Peat, F. David. *Synchronicity: The Bridge Between Matter and Mind*. New York: Bantam Books, 1987.

Tarnas, Richard. *Cosmos and Psyche: Intimations of a New World View*. New York: Ballantine, forthcoming.

Weinberg, Steven. "Lonely planet." *Science and Spirit* 10:1 (April–May 1999).

CHAPTER 2

Bekenstein, Jacob D. "Information in the holographic universe." *Scientific American* (August 2003).

Everett, Hugh. *Rev. Mod. Physics* 29 (1957).

Susskind, Leonard. "A universe like no other." *New Scientist* (1 November 2003).

CHAPTERS 3 AND 5

Cosmology

Bucher, Martin A., Alfred S. Goldhaber, and Neil Turok. "Open universe from inflation." *Physical Review D* 52:6 (15 September 1995).

Bucher, Martin A., and David N. Spergel. "Inflation in a Low-Density Universe." *Scientific American* (January 1999).

Chaboyer, Brian, Pierre Demarque, Peter J. Kernan, and Lawrence M. Krauss. "The age of globular clusters in light of Hipparchos: resolving the age problem?" *Astrophysical Journal* 494:1 (10 February 1998).

Chown, Marcus. "Quantum rebel." *New Scientist* (24 July 2004).

Gribbin, John. *In the Beginning: The Birth of the Living Universe.* New York: Little, Brown & Co., 1993.

Guth, Alan H. *The Inflationary Universe: The Quest for a New Theory of Cosmic Origins.* New York: Perseus Books, 1997.

Hogan, Craig J. *The Little Book of the Big Bang.* New York: Springer Verlag, 1998.

Kafatos, Menas. "Non-locality, foundational principles and consciousness." *The Noetic Journal* 5:2 (January 1999).

Kafatos, Menas, and Robert Nadeau. *The Conscious Universe: Part and Whole in Modern Physical Theory.* New York: Springer Verlag, 1990, 1999.

Krauss, Lawrence M. "The end of the age problem and the case for a cosmological constant revisited." *Astrophysical Journal* 501:2 (10 July 1998).

——. "Cosmological antigravity." *Scientific American* (January 1999).

Leslie, John. *Universes.* London and New York: Routledge, 1989.

——, ed. *Physical Cosmology and Philosophy.* New York: Macmillan, 1990.

Mallove, Eugene F. "The self-reproducing universe." *Sky & Telescope* 76:3 (September 1988).

Michelson, A. A. "The relative motion of the Earth and the luminiferous ether." *American Journal of Science* 22 (1881).

Peebles, P., and E. James. *Principles of Physical Cosmology.* Princeton, N.J.: Princeton University Press, 1993.

Perlmuter, S., G. M. Aldering, M. Della Valle, et al. "Discovery of a supernova explosion at half the age of the universe." *Nature* 391 (1 January 1998).

Prigogine, I., J. Geheniau, E. Gunzig, and P. Nardone. "Thermodynamics of Cosmological Matter Creation." *Proceedings of the National Academy of Sciences USA,* 85 (1988).

Riess, Adam G., Alexei V. Filippenko, Peter Challis, et al. "Observational evidence from supernovae for an accelerating universe and a cosmological constant." *Astronomical Journal* 116:3 (September 1998).

Rees, Martin. *Before the Beginning: Our Universe and Others.* New York: Addison-Wesley, 1997.

Quantum Physics

Barrett, M. D., et al. "Deterministic quantum teleportation of atomic qubits." *Nature* 429 (17 June 2004).

Bohm, David. *Wholeness and the Implicate Order.* London: Routledge & Kegan Paul, 1980.

Buks, E., R. Schuster, M. Heiblum, D. Mahalu, and V. Umansky. "Dephasing in electron interference by a 'which-path' detector." *Nature* 391 (26 February 1998).

Coyle, Michael J. "Localized reduction of the primary field of consciousness as dynamic crystalline states." *The Noetic Journal* 3:3 (July 2002).

Dürr, S., T. Nonn and G. Rempe. "Origin of quantum-mechanical complementarity probed by a 'which-way' experiment in an atom interferometer." *Nature* 395 (3 September 1998).

Gazdag, László. "Superfluid mediums, vacuum spaces." *Speculations in Science and Technology* 12:1 (1989).

Haisch, Bernhard, Alfonso Rueda, and H. E. Puthoff. "Inertia as a zero-point-field Lorentz force." *Physical Review A* 49.2 (1994).

Haroche, Serge. "Entanglement, decoherence and the quantum/classical boundary." *Physics Today* (July 1998).

Herbert, Nick. *Quantum Reality.* Garden City, N.Y.: Anchor-Doubleday, 1987.

Kaivarainen, Alex. "Unified model of bivacuum, particles duality, electromagnetism, gravitation and time: The superfluous energy of asymmetric bivacuum." *The Journal of Non-Locality and Remote Mental Interactions* 1:3 (October 2002).

LaViolette, Paul. *Subquantum Kinetics: A Systems Approach to Physics and Cosmology*. Alexandria, Va.: Starlane Publications, 1994, 2003.

Mueller, Hartmut. "Global scaling—die globale Zeitwelle." *Raum & Zeit,* 19:5, no. 107 (2000).

Riebe, M., et al. "Deterministic quantum teleportation with atoms." *Nature* 429 (17 June 2004).

Rothman, Tony, and George Sudarshan. *Doubt and Certainty*. Reading, Mass.: Perseus Books, 1998.

Sarkadi, Dezső, and László Bodonyi. "Gravity between commensurable masses." Private Research Centre of Fundamental Physics, *Magyar Energetika* 7:2 (1999).

Stapp, Henry P. "Quantum physics and the physicist's view of nature." In *The World View of Contemporary Physics,* edited by Richard E. Kitchener. Albany: State University of New York Press, 1988.

Tittel, W., J. Brendel, H. Zbinden, and N. Gisin. *Phys. Rev. Lett.* 81, 3563 (1998).

Tzoref, Judah. "Vacuum kinematics: a hypothesis." *Frontier Perspectives* 7:2 (1998).

Wagner, E. O. "Structure in the Vacuum." *Frontier Perspectives* 10:2 (1999).

Biology

Behe, Michael J. *Darwin's Black Box: The Biochemical Challenge to Evolution*. New York: Touchstone Books, 1998.

Bischof, Marco. "Introduction to integrative biophysics." In *Lecture Notes in Biophysics,* edited by Fritz-Albert Popp and Lev V. Beloussov. Dordrecht, Netherlands: Kluwer Academic Publishers.

Capra, Fritjof. *The Web of Life: A New Scientific Understanding of Living Systems*. New York: Doubleday, 1996.

Del Giudice, E., G. S. Doglia, M. Milani, C. W. Smith, and G. Vitiello. "Magnetic flux quantization and Josephson behavior in living systems." *Physica Scripta* 40 (1989).

Dobzhansky, Theodosius. *Genetics and the Origin of Species*. 2nd ed. New York: Columbia University Press, 1982.

Dürr, Hans-Peter. "Sheldrake's ideas from the perspective of modern physics." *Frontier Perspectives* 12:1 (Spring 2003).

Eldredge, Niles. *Time Frames: The Rethinking of Darwinian Evolution and the Theory of Punctuated Equilibria.* New York: Simon & Schuster, 1985.

Eldredge, Niles, and Stephen J. Gould. "Punctuated equilibria: an alternative to phylogenetic gradualism." In *Models in Paleobiology,* edited by Thomas J. M. Schopf. San Francisco: Freeman, Cooper, 1972.

Gould, Stephen J. "Irrelevance, submission and partnership: the changing role of paleontology in Darwin's three centennials, and a modest proposal for macroevolution." In *Evolution from Molecules to Men,* edited by D. Bendall. Cambridge: Cambridge University Press, 1983.

Gould, Stephen J., and Niles Eldredge. "Punctuated equilibria: the tempo and mode of evolution reconsidered." *Paleobiology* 3 (1977).

———. "Opus 200." *Natural History* (August 1991).

Goodwin, Brian. "Development and evolution." *Journal of Theoretical Biology* 97 (1982).

———. "Organisms and minds as organic forms." *Leonardo* 22:1 (1989).

———. "On morphogenetic fields." *Theoria to Theory* 13 (1979).

Ho, Mae-Wan. *The Rainbow and the Worm: The Physics of Organisms.* Singapore and London: World Scientific, 1993.

———. "Bioenergetics, biocommunication and organic space-time." In *Living Computers,* edited by A. M. Fedorec and P. J. Marcer. U.K.: The University of Greenwich, March 1996.

———, and Peter Saunders. "Liquid crystalline mesophases in living organisms." In *Bioelectromagnetism and Biocommunication,* edited by M. W. Ho, F. A. Popp, and U. Warnke. Singapore and London: World Scientific, 1994.

Hoyle, Fred. *The Intelligent Universe.* London: Michael Joseph, 1983.

Lieber, Michael M. "Environmentally responsive mutator systems: toward a unifying perspective." *Rivista di Biologia/Biology Forum,* 91:3 (1998).

———. "Hypermutation as a means to globally restabilize the genome following environmental stress." *Mutation Research, Fundamental and Molecular Mechanisms of Mutagenesis* 421:2 (1998).

———. "Force and genomic change." *Frontier Perspectives* 10:1 (2001).

Lorenz, Konrad. *The Waning of Humaneness.* Boston: Little, Brown & Co., 1987.

Maniotis, A., et al. "Demonstration of mechanical connections between integrins, cytoskeletal filaments, and nucleoplasm that stabilize nuclear structure." *Proceedings of the National Academy of Sciences USA,* 4:3 (1997).

Primas, Hans, H. Atmanspacher, and A. Amman. *Quanta, Mind and Matter: Hans Primas in Context.* Dordrecht, Netherlands: Kluwer, 1999.

Senapathy, Periannan. *Independent Birth of Organisms.* Madison, Wis.: Genome Press, 1994.

Smith, Cyril W. "Is a living system a macroscopic quantum system? *Frontier Perspectives* (Fall/Winter 1998).

Sitko, S. P., and V. V. Gizhko. "Towards a quantum physics of the living state." *Journal of Biological Physics* 18 (1991).

Steele, Edward J., R. A. Lindley, and R. V. Blandon. *Lamarck's Signature: New Retro-genes Are Changing Darwin's Natural Selection Paradigm.* London: Allen & Unwin, 1998.

Taylor, R. "A gentle introduction to quantum biology." *Consciousness and Physical Reality* 1:1 (1998).

Waddington, Conrad. "Fields and gradients." In *Major Problems in Developmental Biology,* edited by Michael Locke. New York: Academic Press, 1966.

Welch, G. R. "An analogical 'field' construct in cellular biophysics: history and present status." *Progress in Biophysics and Molecular Biology* 57 (1992).

——, and H. A. Smith. "On the field structure of metabolic space-time." In *Molecular and Biological Physics of Living Systems,* edited by R. K. Mishra. Dordrecht, Netherlands: Kluwer, 1990.

Wolkowski, Z. W. "Recent advances in the phoron concept: an attempt to decrease the incompleteness of scientific exploration of living systems." In *Biophotonics—Nonequilibrium and Coherent Systems in Biology, Biophysics and Biotechnology,* edited by L. V. Beloussov and F. A. Popp. Moscow: Moscow University Press, 1995.

Zeiger, Berndt F., and Marco Bischof. "The quantum vacuum and its significance in biology." *Proceedings of the Third International Hombroich Symposium on Biophysics,* Neuss, Germany (1998).

Consciousness Research

Backster, Cleve. "Evidence of a primary perception in plant life." *International Journal of Parapsychology* 10:4 (1968).

——. "Evidence for a primary perception at the cellular level in plants and animals." American Association for the Advancement of Science, Annual Meeting 26–31 (January 1975).

——. "Biocommunications capability: Human donors and *in vitro* leukocytes." *International Journal of Biosocial Research* 7:2 (1985).

Benor, Daniel J. *Healing Research: Holistic Energy Medicine and Spiritual Healing.* Vol. 1. London: Helix Editions, 1993.

——. "Survey of spiritual healing research." *Contemporary Medical Research* 4:9 (1990).

Bischof, Marco. "Holism and field theories in biology—non-molecular approaches and their relevance to biophysics." In *Biophotons,* edited by J. J. Chang, J. Fisch, and F. A. Popp. Dordrecht, Netherlands: Kluwer, 1998.

——. "Field concepts and the emergence of a holistic biophysics." In *Biophotonics and Coherent Structures,* edited by L. V. Beloussov, V. L. Voeikov, and R. Van Wijk. Moscow: Moscow University Press, 2000.

Bohm, David. Interview by John Briggs and F. David Peat. *Omni* 9:4 (1987).

——, and Basil Hiley. *The Undivided Universe.* London: Routledge, 1993.

Braud, W. G. "Human interconnectedness: research indications." Revision 14:3 (1992).

——, and M. Schlitz. "Psychokinetic influence on electrodermal activity." *Journal of Parapsychology* 47 (1983).

Byrd, R. C. "Positive therapeutic effects of intercessory prayer in a coronary care population." *Southern Medical Journal* 81:7 (1988).

Cardeña, Etzel, Steven Jay Lynn, and Stanley Krippner. "Varieties of anomalous experience: Examining the scientific evidence." American Psychological Association, Washington, D.C., 2000.

Dossey, Larry. *Recovering the Soul: A Scientific and Spiritual Search.* New York: Bantam, 1989.

——. *Healing Words: The Power of Prayer and the Practice of Medicine.* San Francisco: Harper 1993.

——. "Era III medicine: the next frontier." *ReVision* 14:3 (1992).

Elkin, A. P. *The Australian Aborigines.* Sydney: Angus & Robertson, 1942.

——, and Stanley Krippner. *The Mythic Path.* New York: Tarcher Putnam, 1997.

Freeman, W. J. and J. M. Barrie. "Chaotic oscillations and the genesis of meaning in cerebral cortex." In *Temporal Coding in the Brain,* edited by G. Bizsaki. Berlin: Springer Verlag, 1994.

Grinberg-Zylberbaum, Jacobo, M. Delaflor, M. E. Sanchez-Arellano, M. A. Guevara, and M. Perez. "Human communication and the electro-physiological activity of the brain." *Subtle Energies* 3:3 (1993).

Grof, Stanislav. *The Adventure of Self-discovery.* Albany: State University of New York Press, 1988.

——. "Healing and heuristic potential of non-ordinary states of conscious-ness: Observations from modern consciousness research." *Mimeo,* 1996.

——, with Hal Zina Bennett. *The Holotropic Mind.* San Francisco: Harper, 1993.

Hansen, G. M., M. Schlitz, and C. Tart. "Summary of remote viewing research." In Russell Targ and K. Harary, *The Mind Race.* New York: Villard, 1984.

Honorton, C., R. Berger, M. Varvoglis, M. Quant, P. Derr, E. Schechter, and D. Ferrari. "Psi-communication in the Ganzfeld: Experiments with an automated testing system and a comparison with a meta-analysis of earlier studies." *Journal of Parapsychology* 54 (1990).

Keen, Jeffrey S. "Mind-created dowsable fields." *Dowsing Research Group: The First 10 Years.* Wolverhampton, U.K.: Magdalena Press, 2003.

Montecucco, N. "Cyber: Ricerche Olistiche (Nitamo Montecucco)." *Cyber* (Milan) (November 1992).

Morgan, Marlo. *Mutant Message Down Under.* New York: HarperCollins, 1991.

Nelson, John E. *Healing the Split.* Albany: State University of New York Press, 1994.

Persinger, M. A., and S. Krippner. "Dream ESP experiments and geomag-netic activity." *The Journal of the American Society for Psychical Research* 83, 1989.

Playfair, Guy. *Twin Telepathy: The Psychic Connection.* London: Vega Books, 2002.

Puthoff, Harold, and Russell Targ. "A perceptual channel for information transfer over kilometer distances: historical perspective and recent research." *Proceedings of the IEEE* 64 (1976).

Rein, Glen. "Biological effects of quantum fields and their role in the natural healing process." *Frontier Perspectives* (Fall/Winter 1998).

———. "Biological interactions with scalar energy-cellular mechanisms of action." *Proceedings of the 7th International Association of Psychotronics Research Conference.* Atlanta, Georgia (December 1988).

Ring, Kenneth. *Life at Death: A Scientific Investigation of the Near-Death Experience.* New York: Coward, McCann and Geoghegan, 1980.

———. *Heading Toward Omega: In Search of the Meaning of the Near-Death Experience.* New York: Morrow, 1984.

———. "Near-death and out-of-body experiences in the blind: A study of apparent eyeless vision." *Journal of Near-Death Studies* 16:2 (Winter 1997).

Rosenthal, R. "Combining results of independent studies." *Psychological Bulletin* 85 (1978).

Sági, Maria. "Holistic healing as fresh evidence for collective consciousness." *World Futures* 48 (1997).

———. "Healing through the QVI-field." In *The Evolutionary Outrider,* edited by David Loye. London: Adamantine Press, 1998.

Smith, Cyril W. "Is a living system a macroscopic quantum system?" *Frontier Perspectives* (Fall/Winter 1998).

Targ, Russell, and Harold A. Puthoff. "Information transmission under conditions of sensory shielding." *Nature* 251 (1974).

Targ, Russell, and K. Harary. *The Mind Race.* New York: Villard Books, 1984.

Tart, Charles. *States of Consciousness.* New York: Dutton, 1975.

Thalbourne, M. A. and P. S. Delin. "A common threat underlying belief in the paranormal, creative personality, mystical experience and psychopathology." *Journal of Parapsychology* 58 (1994).

———. "Transliminality: its relation to dream life, religiosity and mystical experience." *International Journal for the Psychology of Religion* 9 (1999).

Tiller, William A. "Subtle energies in energy medicine." *Frontier Perspectives* 4:2 (Spring 1995).

Ullman, M., and S. Krippner. *Dream Studies and Telepathy: An Experimental Approach*. New York: Parapsychology Foundation, 1970.

Varvoglis, Mario. "Goal-directed- and observer-dependent PK: An evaluation of the conformance-behavior model and the observation theories." *The Journal of the American Society for Psychical Research* 80 (1986).

CHAPTER 4

Akimov, A. E., and G. I. Shipov. "Torsion fields and their experimental manifestations." *Journal of New Energy* 2:2 (1997).

——, and V. Ya. Tarasenko. "Models of polarized states of the physical vacuum and torsion fields." *Soviet Physics Journal* 35:3 (1992).

Gazdag, László. *Beyond the Theory of Relativity*. Budapest: Robottechnika Kft., 1998.

——. "Superfluid mediums, vacuum spaces." *Speculations in Science and Technology* 12:1 (1989).

——. "Combining of the gravitational and electromagnetic fields." *Speculations in Science and Technology* 16:1 (1993).

Grof, Stanislav, with Hal Zina Bennett. *The Holotropic Mind*. San Francisco: Harper, 1993.

Haisch, Bernhard, Alfonso Rueda, and H. E. Puthoff. "Inertia as a zero-point-field Lorentz force." *Physical Review A* 49:2 (1994).

James, William, cited in M. Ferrari, "William James and the denial of death." *Journal of Consciousness Studies* (2002).

Mitchell, Edgar R. *The Way of the Explorer: An Apollo Astronaut's Journey through the Material and Mystical Worlds*. New York: Putnam, 1996.

Puthoff, Harold. "Ground state of hydrogen as a zero-point-fluctuation-determined state." *Physical Review D* 35:10 (1987).

——. "Source of vacuum electromagnetic zero-point energy." *Physical Review A* 40:9 (1989).

——. "Gravity as a zero-point-fluctuation force." *Physical Review A* 39:5 (1989, 1993).

Rueda Alfonso, and Bernhard Haisch. "Inertia as reaction of the vacuum to accelerated motion." *Physics Letters A* 240 (1998).

Sakharov, A. "Vacuum quantum fluctuations in curved space and the theory of gravitation." *Soviet Physics-Doklady,* 12:11 (1968).

Schwarzschild, B. "Very distant supernovae suggest that the cosmic expansion is speeding up." *Physics Today* 51:6 (1998).

Shipov, G. I. *A Theory of the Physical Vacuum: A New Paradigm.* Moscow: International Institute for Theoretical and Applied Physics RANS, 1998.

Stevenson, Ian. *Children Who Remember Previous Lives.* Charlottesville: University Press of Virginia, 1987.

——. *Cases of the Reincarnation Type.* 4 vols., Charlottesville: University Press of Virginia, 1975–83.

——. *Reincarnation and Biology: A Contribution to the Etiology of Birthmarks and Birth Defects.* 2 vols. Westport, Conn.: Praeger, 1997.

Stapp, Henry P. "Harnessing science and religion: implications of the new scientific conception of human beings." *Research News and Opportunities in Science and Theology* 2001:1 (February 2001).

Tiller, William. "Towards a predictive model of subtle domain connections to the physical domain aspect of reality: the origins of wave-particle duality, electric-magnetic monopoles and the mirror principle." *Journal of Scientific Exploration* 13:1 (1999).

CHAPTER 6

Susskind, Leonard. "A universe like no other." *New Scientist* (1 November 2003).

CHAPTERS 7 AND 8

Aczel, Amir D. *Probability. Why There Must Be Intelligent Life in the Universe.* New York: Harcourt Brace, 1998.

Aurobindo, Sri. *The Life Divine.* 2nd printing. New York: Sri Aurobindo Library, 1951.

Bailey, Alice. *Telepathy and the Etheric Vehicle.* New York: Lucis, 1950.

Barrow John D., and Frank J. Tipler. *The Anthropic Cosmological Principle.* London and New York: Oxford University Press, 1986.

Beck, Don, and Christopher C. Cowan. *Spiral Dynamics: Mastering*

Values, Leadership and Change. Oxford: Blackwell, 1996.

Botkin, Allan, and R. Craig Hogan. *Reconnections: The Induction of After-Death Communication in Clinical Practice.* Charlottesville, Va.: Hampton Roads, forthcoming (cited from MS of January 10, 2004).

Chalmers, David J. *The Conscious Mind: In Search of a Theory of Conscious Experience.* New York: Oxford Press, 1996.

——. "The puzzle of conscious experience." *Scientific American* 273:6 (1995).

Dawkins, Richard. *The Blind Watchmaker.* London: Longmans, 1986.

Drake, Frank. *Intelligent Life in Space.* New York: Macmillan, 1964.

Dyson, Freeman. *Infinite in All Directions.* New York: Harper & Row, 1988.

Eddington, Sir Arthur. "Defense of mysticism." In Ken Wilber, *Quantum Questions: Mystical Writings of the World's Great Physicists.* Boston: Shambhala, 1984.

Fechner, Gustav. Quoted in William James, *The Pluralistic Universe.* London, New York, and Bombay: Longmans, Green & Co., 1909.

Fodor, Jerry A. "The big idea." *New York Times Literary Supplement,* July 3, 1992.

Gebser, Jean. *Ursprung und Gegenwart.* Stuttgart: Deutsche Verlagsanstalt, 1949.

Grof, Stanislav. *The Cosmic Game. Explorations at the Frontiers of Human Consciousness.* Albany: State University of New York Press, 1999.

Haldane, J. B. S. "The origin of life." *Rationalist Annual* 148 (1928).

Huang, Su-Shu. *American Scientist* 47 (1959).

Oparin, A. I. *Origin of Life.* New York: Dover, 1938.

Ponnamperuma, Cyril. "Experimental studies on the origin of life." *Journal of the British Interplanetary Society* 42 (1989).

——. "The origin, evolution, and distribution of life in the universe." *Cosmic Beginnings and Human Ends,* edited by Clifford N. Matthews and Roy A. Varghese. Chicago and La Salle: Open Court, 1995.

Pribram, Karl H. "Consciousness reassessed." *Mind and Matter* (2004).

Russell, Bertrand. "A freeman's worship." In *The Basic Writings of Bertrand Russell 1903–1959,* edited by R. E. Egner and L. D. Dennon. New York: Simon & Schuster, 1960.

Russell, Peter. *The White Hole in Time* (reissued as *Waking Up in Time*). Novato, Calif.: Origin Press, 1992, 1998.

Sagan, Carl. *Intelligent Life in the Universe.* New York: Emerson Adams Press, 1966.

Schrödinger, Erwin. *My View of the World.* Cambridge: Cambridge University Press, 1964.

Shapley, Harlow. *Of Stars and Men.* Boston: Beacon, 1958.

Stapp, Henry P. "Harnessing science and religion: implications of the new scientific conception of human beings." *Research News and Opportunities in Science and Theology* 2001:1 (February 2001).

Taormina, Robert J. "A New Consciousness for Global Peace." *Proceedings, Third International Symposium on the Culture of Peace,* Baden Baden, 1999.

Teilhard de Chardin, Pierre. *The Future of Man.* 1959. Reprint New York: Harper & Row, 1964.

Ward, Peter B. *Rare Earth: Two Tiers of Life in the Universe.* New York: Springer Verlag, 2000.

Wilber, Ken. *Up from Eden: A Transpersonal View of Human Evolution.* Boulder: Shambhala, 1983.

——. "Physics, mysticism, and the new holographic paradigm." In *The Holographic Paradigm,* edited by K. Wilber, Boston: Shambhala, 1982.

——. "An integral theory of consciousness." *Journal of Consciousness Studies* 4:1 (1997).

——. *A Theory of Everything: An Integral Vision for Business, Politics, Science and Spirituality.* Boston: Shambhala, 2000.

Index

BOOKS OF RELATED INTEREST

A New Science of Life
The Hypothesis of Morphic Resonance
by Rupert Sheldrake

Matrix of Creation
Sacred Geometry in the Realm of the Planets
by Richard Heath

Genesis of the Cosmos
The Ancient Science of Continuous Creation
by Paul A. LaViolette, Ph.D.

Alternative Science
Challenging the Myths of the Scientific Establishment
by Richard Milton

Shattering the Myths of Darwinism
by Richard Milton

Stalking the Wild Pendulum
On the Mechanics of Consciousness
by Itzhak Bentov

The Universe Is a Green Dragon
A Cosmic Creation Story
by Brian Swimme, Ph.D.

The Biology of Transcendence
A Blueprint of the Human Spirit
by Joseph Chilton Pearce

Inner Traditions • Bear & Company
P.O. Box 388
Rochester, VT 05767
1-800-246-8648
www.InnerTraditions.com

Or contact your local bookseller

SSN

SSN

457-58-7951